RAD FUTURE

RAD FUTURE

THE UNTOLD STORY OF NUCLEAR ELECTRICITY AND HOW IT WILL SAVE THE WORLD

ISABELLE BOEMEKE

THESIS

Thesis
An imprint of Penguin Random House LLC
1745 Broadway, New York, NY 10019
penguinrandomhouse.com

Copyright © 2025 by Isabelle Boemeke

Penguin Random House values and supports copyright. Copyright fuels creativity, encourages diverse voices, promotes free speech, and creates a vibrant culture. Thank you for buying an authorized edition of this book and for complying with copyright laws by not reproducing, scanning, or distributing any part of it in any form without permission. You are supporting writers and allowing Penguin Random House to continue to publish books for every reader. Please note that no part of this book may be used or reproduced in any manner for the purpose of training artificial intelligence technologies or systems.

Thesis with colophon is a trademark of Penguin Random House LLC.

Most Thesis books are available at a discount when purchased in quantity for sales promotions or corporate use. Special editions, which include personalized covers, excerpts, and corporate imprints, can be created when purchased in large quantities. For more information, please call (212) 572-2232 or e-mail specialmarkets@penguinrandomhouse.com. Your local bookstore can also assist with discounted bulk purchases using the Penguin Random House corporate Business-to-Business program. For assistance in locating a participating retailer, e-mail B2B@penguinrandomhouse.com.

Illustrations drawn by Maria Nogueira Nössing

Book design by Nicole LaRoche

Library of Congress Control Number: 2025002158
ISBN 9780593716328 (hardcover)
ISBN 9780593716335 (ebook)

Printed in the United States of America
1st Printing

The authorized representative in the EU for product safety and compliance is
Penguin Random House Ireland, Morrison Chambers, 32 Nassau Street,
Dublin D02 YH68, Ireland, https://eu-contact.penguin.ie.

For Joe, who said "That's a great idea" when I first pitched my wild plan to become a nuclear energy influencer. This book wouldn't exist if it weren't for you.

CONTENTS

INTRODUCTION — xiii

PART 1: A WHOLE NEW ERA

1. **THE DAWN OF THE ATOMIC AGE** — 3
   - What Could Have Been — 5
   - What Actually Happened — 8
   - The Impact — 10
   - Infinite Possibility — 11

PART 2: ALCHEMY

2. **HUMANS AND ENERGY** — 19
   - Energy Is the Currency of the Universe — 19
   - Energy Timeline — 22

3. **FOSSIL FUELS' FINAL BOSS** — 29
   - How Nuclear Reactors Work — 31
   - Nuclear Is RAD(IANT) — 39
   - Nuclear Is Natural — 53
   - Nuclear Has the Least Impact — 54

4. DISPELLING MYTHS — 59
- What About Chernobyl? — 61
- What About Radiation? — 68
- What About Uranium Mining? — 74
- Mining Sucks — 77
- What About Tritium? — 81
- What About Nuclear Waste? — 84

5. DEGROWTH — 96
- The Problem with Degrowth — 97
- We Will Need a Lot of Electricity — 102

6. CLEAN ENERGY REVOLUTION — 108
- 100 Percent rEnEwAbLe — 108
- Clean Energy — 112
- The Ultimate Energy Diet — 117

PART 3: DIPLOMACY: FIGHTING A WORLD AGAINST NUCLEAR

7. CHANGING TIDES — 133
- Nuclear Before Nukes — 133
- Atomic Fervor — 137
- Mixed Feelings — 141
- The Antinuclear Movement — 143
- Grubby Hands — 151

8. THREE DECADES OF BAD ENERGY POLICY — 158
- A Tale of Two Decarbonizations — 158
- The Land of Wine, Great Aesthetics, and Nuclear Electricity — 163
- Who Killed Nuclear Electricity? — 165

9. VIBE SHIFT	171
Gender Gap	185
The Nuclear Renaissance	188
Small Modular Reactors	193
10. MANIFESTING A RAD FUTURE	199
How You Can Help	203
ACKNOWLEDGMENTS	209
GLOSSARY	213
NOTES	219
ILLUSTRATION CREDITS	231
INDEX	235

Nothing in life is to be feared. It is only to be understood. Now is the time to understand more, so that we may fear less.

—Marie Curie

INTRODUCTION

Half-awake and still squinting, I opened Twitter (now known as X) expecting the usual—some memes, a random celebrity getting canceled, maybe a few questionable hot takes. Instead, I saw my home on fire. My homeland of Brazil was burning, and there I was seeing it all unfold on a tiny screen from the comfort of my bed in California, thousands of miles away.

It was 2019, and the Amazon rainforest was ablaze. The hashtag #PrayForAmazonas topped Twitter; satellites had identified over forty thousand fires raging in the region.

This wasn't the first time fires in the Amazon left it looking like a scene from a postapocalyptic movie. But in 2019, at age twenty-nine, the images hit me differently. Probably because I was still shaken by seeing equally hellish images from the other side of the planet. Earlier that year, Australia had experienced one of the worst series of bushfires in recorded history. Walls of flame torched the country during that "Black Summer," burning koalas and killing humans.

The consensus among experts was that at the root of it, climate change was to blame for the escalation in fires that claimed huge swaths of once-pristine forest or bush.

Like everyone in my generation, I grew up hearing—and worrying—about climate change. But tbh, it always seemed like a problem for... someone else. Those people over there. An issue some future someone would solve for us. I mean, they're patching up the ozone layer. For sure someone would figure this one out too, right?

Those images of the Amazon burning were a brutal wake-up call—a clear sign that climate change had turned into a bigger crisis. What once felt like an abstract issue suddenly filled me with despair about the state of the world and our lack of progress in solving it.

Brazilians feel a deep sense of pride in the Amazon and a responsibility to protect it. Beyond its unmatched biodiversity and its role as a home to Indigenous communities, the forest is often called the "lungs of the world." This massive breathing ecosystem absorbs CO_2 and releases oxygen, helping to keep the planet in balance.

It's true that tracing specific fires or hurricanes directly to climate change is basically impossible. In the Amazon, most fires are started by people illegally clearing land for agriculture, logging, or mining. But as the planet gets hotter and drier, these fires are becoming more frequent and severe—and it's real people who are paying the price, watching their homes and land go up in smoke.

Growing up in Rio Grande do Sul, Brazil's southernmost state known for the grassy Pampas Plain, I felt that climate change, though present, was more like background noise. As I got older, the mood began to shift, and by my mid-twenties, a sense of doom about the planet and our future had become a constant presence.

Everyone seemed to be echoing some version of the same bleak thought: Humans are a plague on Earth. A virus. We destroy everything we touch, and Earth would be better off if we just disappeared. Dark.

While all major religions highlight the struggle between good and evil in human hearts, the view that people are evil while nature is good is quite modern. It became mainstream in the 1970s as we started waking up to the impacts of human activities on the environment. That awakening was incredibly valuable and inspired the creation of the environmental movement, which has had important wins like the preservation of wilderness habitats and cleaning up polluted air and water. But painting our entire species as evil is simplistic and ultimately leads to some pretty disturbing beliefs, like thinking that humans should go extinct. It's cringe to admit but I had fully bought into that way of thinking, like so many of those around me. Our conversations turned to how irresponsible it was to have kids, because of their carbon footprints. As if humans were leeches whose only contribution was more CO_2.

Despite how messed up this people-hating nonsense is, it has stuck. Watch any film by beloved nature documentarian David Attenborough and the underlying message is always some flavor of "Humans suck." Even iconic physicist Stephen Hawking once said humans are "chemical scum." In 2020, during the COVID-19 pandemic, the hashtag #HumansAreTheVirus went viral. The sentiment was that the pandemic was the punishment we deserved for destroying our planet.

More depressing than the antihuman rhetoric is the pervasive vibe of hopelessness in the air. This pessimism is crushing, especially for younger people. I've spoken to college students who believe climate change will kill everyone in a few decades. They try not to think too much about their future because it fills them with dread and anxiety. I used to feel the same way. But as a reformed human-shading pessimist, I have good news: There's a way out of this dark hole of negativity and helplessness. For me, the light at the event horizon came when I read a book by British physicist David Deutsch called *The Beginning of Infinity*. He makes the case that humans aren't bad or insignificant. According to Deutsch, we might be the coolest thing in the entire universe. That's obviously true only if there aren't any space-traveling aliens out there (which seems unlikely, but that's a story for another book).

Deutsch believes humans are special because of our unique ability to create knowledge and use it to solve problems and transform the world around us. And here's the really cool part: He argues that *every* problem is solvable, provided we have enough information about it.

Framing the universe as a box filled with problems that we *can* and *get* to solve has been a game changer for me. It meant I stopped freezing when facing the world's seemingly endless issues and started seeing them as opportunities instead. Suddenly, my despair was replaced by motivation. Empowered by this problem-solving mindset, I went from viewing climate change as an inescapable death sentence to just another very complex, but ultimately solvable, problem.

Adopting this mindset didn't just change how I viewed problems; it also transformed how I saw humanity itself. I became shockingly aware of how our culture pushes the view of humans as destroyers—destroyers of others, of peace, of nature. I prefer to see us as creators—creators of love, of art, of knowledge, and of the future.

That's why, when I saw pictures and videos of the world on fire, after a brief moment of despair, I started thinking about how I could be a part of the solution.

Looking back now, I know that sense of conviction seems sweet and even naive. How could *I* contribute to addressing one of the most challenging issues our civilization is facing? I had no blueprint, no grand strategy, just the desire to make a difference.

It turns out a little seed had already been planted in my mind several years earlier...

I was a nineties baby raised in a very small rural town in the south of Brazil. When people think of Brazil, they picture a tropical climate with beautiful sunny beaches or the lush Amazon rainforest. But where I am from is mostly farmland, and we get the pleasure of both extreme heat in the summers *and* frigid cold in the winters. Almost nobody had air-conditioning or central heating. I have clear memories of freezing during the winter months, wearing my coat inside the house and sipping maté, desperately trying to stay warm.

At the age of sixteen, I was approached by a model scout as I was walking out of school, changing my life overnight. He convinced me to join a national modeling competition, a decision that led me away from university and into the fast-paced and mildly sketchy world of fashion. That career took me all over the globe from a young age and eventually it took me to the United States of America.

In 2009, while living in Miami, I stumbled upon a book that would become one of the most important in my life: *The Greatest Show on Earth* by evolutionary biologist Richard Dawkins. It introduced me to evolution, a topic my Catholic-school education had kind of ignored, and it sparked an obsession with science and learning. I soon began reading as many popular science books as I could, but my newfound passion felt totally out of place in the fashion industry, where most people didn't share that interest. I was desperate to find a community where I could engage with like-minded folks, so I turned to Twitter and started following a bunch of scientists. Among them was American planetary scientist Carolyn Porco.

Years later, in 2015, a post by Porco caught my attention and unknowingly set me on a whole new path—one that brought me here, typing the very words you're reading in this book. In her post, she mentioned something called "molten salt thorium reactors." You're probably thinking, "Molten what?" And that was exactly my reaction back then. But here was an icon of science, a woman whose career revolved around pursuing the truth, speaking positively about nuclear electricity.* I thought nuclear was supposed to be the bad guy in our energy saga.

At the time, I knew nothing about nuclear fission, let alone whatever the hell a molten salt thorium reactor was supposed to be. But I started asking people what they thought about nuclear and I was surprised by their very strange responses. Whispering behind closed doors, many admitted that nuclear is safe. In fact, most even said it's essential for combating climate change. Yet, almost always, they would quickly add, "But people hate it."

Those responses stuck with me. How could a technology that is both safe and necessary be so hated?

When I decided to dedicate my time to tackling the climate crisis, nuclear electricity kept popping up on my radar. The more I

---

* You might have heard about "nuclear energy" and wonder why I am calling it "nuclear electricity" instead. The term "nuclear energy" has a lot of baggage. Calling it "nuclear electricity" highlights the use of this technology I am advocating for: making electricity.

learned, the more obvious it became that it was essential for winning this fight. The deeper I dove into its science and history, the more fascinated I became by the massive gap between public perception and reality. Here was one of the safest energy sources, capable of providing reliable, clean electricity while using the least amount of land and resources, yet it was demonized. I was consumed with figuring out how and why this negative image had been created and embedded in our collective consciousness. After months of piecing the puzzle together, it finally clicked.

Most people just want to get home, turn on the lights, and charge their phones without thinking about where that electricity comes from. Unless there's a headline-worthy update, no one pays attention to nuclear reactors.

When reactors make the news, it's almost always because something has gone wrong. The media loves a good disaster, and nuclear plants, when they do mess up, make for wild headlines. "Nuclear Plant Accident!" gets significantly more clicks than "Nuclear Plant Quietly Powers Millions of Homes Without Incident." As a result, people only associate nuclear electricity with accidents, making those events the first thing they picture when they hear the word "nuclear."

Hollywood hasn't helped either. Nearly every TV show or movie that mentions the technology makes it sound scary or straight-up evil. For the older crowd, the word "nuclear" conjures hazy memories of Cold War bomb drills, Chernobyl's ghost town, and barrels of what looks like toxic slime from a nineties cartoon (which is not what nuclear waste looks like, but we'll get into that in chapter 4). Younger generations might not carry all that cultural baggage, but they've still inherited a vague idea that nuclear electricity is bad and dangerous.

This negative image isn't justified, and it didn't happen by mistake. It is the result of decades of relentless campaigning by antinuclear activists and the fossil fuel industry.

Throughout my journey, I realized that nuclear electricity is about a lot more than ditching fossil fuels. It's about entering a whole new era for the human species—one of energy abundance for everyone, while minimizing our impact on the environment.

This realization left me with a tough challenge. How do you tell people that the thing they've been taught to fear is their ticket to a better future? It's like trying to convince someone that the supervillain in their favorite movie was the hero all along.

After a ten-day fast (true story) I had an unusual idea: What if I became a nuclear energy influencer? I wanted to use the platform I had built through my modeling career to talk about something that mattered more than selling lipstick.

Inspired aesthetically by my favorite musician, Rosalía, and AI influencers, I crafted a persona to share my message. I called her ISODOPE, a pun on "isotope" (we'll get into isotopes in chapter 3). ISODOPE content starts with your usual influencer tropes—workout, skincare, or "What I Eat in a Day" videos—and transitions into talking about nuclear electricity instead. You might think it's an insane idea—the type of idea you could only have after spending ten days without food. And you would be right. But it caught on. My videos were viewed by millions and shared by respected industry leaders. They also appealed to younger people who would otherwise never engage with content about this topic. My, let's say, unusual approach to communicating even led me to the TED stage in Vancouver, where I delivered a talk that's been viewed over 1.7 million times.

A little over a year after ISODOPE's entrance into the digital world, I decided it was time to bring the movement into the real

Introduction

world. In December 2021, I set my sights on the iconic Diablo Canyon, California's only nuclear power plant and the birthplace of America's antinuclear movement.

Diablo Canyon was on the brink of a premature shutdown—not because there was anything wrong with it, but because California politicians were caving to pressure from the antinuclear crowd. The problem was that shutting it down would lead to more greenhouse gas emissions, accelerating climate change, and increasing the risk of extreme weather events, rising sea levels, and devastating impacts on ecosystems and human lives.

I couldn't just stand by and let that happen, at least not without a good fight. Even though I didn't have a road map, saving Diablo Canyon became my mission. I launched a grassroots campaign and organized the largest pronuclear rally in U.S. history. There's a whole story behind the rally, including a wild blimp-wrangling adventure, which I'll get into in chapter 9. But for now, here's the spoiler: We won.

On September 1, 2022, nearly nine months after launching the campaign, I sat at my desk, eight months pregnant, and watched in awe as legislation to save Diablo Canyon passed with a 67–3 vote. My optimism about humanity had changed not only the fate of Diablo Canyon, but my whole life. My dread of the future was gone.

Here I was, about to bring more humans into the world.

You might be asking yourself why you should listen to me; after all, I am not a nuclear engineer or an energy policy wonk. I'm not pretending to be. You can consider me a sort of translator—I make complicated and boring stuff easy and fun to understand.

Introduction

I'm not presenting any original science or resorting to obscure references, and every piece of data in this book has been fact-checked by multiple experts. Opinions and sass, on the other hand, are my own.

I'm passionate about building a rad future: a future of radical abundance and human flourishing for all, not just for people in rich countries. A future where electricity is clean and has less impact on the planet and human health. A future where wars aren't funded by our addiction to fossil fuels.

What if I told you that nuclear electricity offers our best shot at making this future a reality?

This book is for those who truly care about finding solutions to the climate crisis and want to become climate change-fighting superheroes. My hope is that by the end, you'll feel inspired by what humans can achieve and excited about the epic future we can create together. Most of all, I hope you'll walk away stoked about nuclear electricity's potential and ready to have informed and productive conversations about it.

The book is divided into three parts and ten chapters. Each chapter is further divided into sections, which are all fairly short. This way, you can jump around to the parts you feel most curious about. At the end of each chapter there's a TL;DR (too long; didn't read) that summarizes the content into its most essential takeaway. In other words, there's no wrong way to read this book—and skimming is encouraged. You don't need to know everything there is to know about nuclear to become a good advocate for it. If you're feeling skeptical, I recommend starting out by reading all the TL;DRs. I promise you'll be so intrigued—and enraged—by how humans have fumbled the bag on this technology that you'll end up circling back to read more.

Introduction

Now that I've introduced you to my whole thing, it's time to introduce you to my good friend, nuclear electricity. Buckle up and let's jump right into this wild journey of badly timed discoveries, well-meaning celebrity smear campaigns, and the evolution of a technology that can still change *everything*.

PART 1
A WHOLE NEW ERA

1: THE DAWN OF THE ATOMIC AGE

The hallways of the stone building are dead silent as fluffy snow falls outside. Most researchers have gone home, but Otto Hahn and Fritz Strassmann are still holed up in their lab, messing with equipment. Fritz looks out the window with his icy blue eyes and zones out for a sec, questioning his choices. A newlywed, he pictures his wife all alone at home, playing a sad tune on the violin. What would he be doing if he weren't a scientist? Hopefully something more fun than spending his nights in a dimly lit room with an old guy and a bunch of beakers.

They have spent all day bombarding uranium—a radioactive element—with neutrons. Lots of physicists, including the famous Enrico Fermi, have done this before. But Otto and Fritz are *chemists*, and their analysis is showing them something new.

Fritz gets up from his hard metal stool and walks across the room, trying to remember what it feels like to have blood flowing to his legs. He's had enough science for the day; he's ready to go home. But just as he is about to hang up his white lab coat and make a break for it, Otto convinces him to run the experiment one more time.

Their lab setup looks like a scene straight out of a B movie about mad scientists, complete with bubbling beakers and humming gadgets. They set up their uranium sample for what feels like the millionth time, hit it with neutrons, and do another chemical analysis of it. Maybe this time they'll be able to figure out what the hell is happening.

According to the current scientific understanding, they should be detecting something heavier than uranium.* But Otto and Fritz are detecting barium, an element that is a lot lighter than uranium. Almost *half* its weight, to be precise . . .

They rally and repeat the experiment a bunch of times, probably hoping the universe will start making sense again. But it doesn't. Each time, they get barium. It's almost like the uranium atom is *splitting* into completely different elements. Wild. By now, it's very late and they're either geniuses or completely unhinged. Fritz and Otto decide to call it a night and write a letter in the morning to their colleague Lise Meitner, a badass physicist, who's in Sweden.

When Lise gets the letter, she's as confused by the results of the experiment as they are. She teams up with her nephew, fellow physicist Otto Frisch, to try to figure this out. They decide to go for a walk in the snow, probably because the best ideas happen when you are freezing. After a few minutes of walking and shivering, they come up with a crazy idea: What if the nucleus of the uranium atom is splitting in half and releasing a lot of energy in the

---

* They thought the U-235 atoms in their uranium sample would absorb the neutrons and become a transuranic element—something new and strange that had a higher atomic weight than uranium, like neptunium or plutonium.

process? After crunching some numbers on scraps of paper, they realize they're onto something huge—like world-changing huge.

Less than a month after their wild night at the lab, Otto and Fritz publish a paper on their observations, but they won't say an atom had split. "As nuclear chemists we cannot bring ourselves to take this step, so contradictory to all the experience of nuclear physics." That seems fair, to be honest. Who wants to go down in history as the guys who cried split atoms over some spilled barium?

But Lise Meitner couldn't care less if others think she is delulu. She has the math to prove it. Lise and her nephew Otto replicate the experiment for themselves and publish the first two official papers on what they dub "nuclear fission." Legendary.

Just imagine how psyched Meitner, Frisch, Hahn, and Strassmann get when they realize their discovery will unlock the powerful energy trapped inside the atom and usher in a whole new era. From now on, the world will never be the same. Except for one detail. The real-life version of this story took place in 1938 . . . in Germany.

WHAT COULD HAVE BEEN

Word of the discovery of nuclear fission gets out quickly. At the University of Chicago in the United States, scientists jump up from their desks and awkwardly dance in their offices, celebrating the news. Some even start crying. They know this technology is a complete game changer—a source of power so dense that it will put an end to humanity's scarcity era. Never again will we struggle to find enough energy to meet the growing demands of society, a problem that has led to competition, wars, and economic drama. Just a couple of years later, that same group of scientists from the University of Chicago finishes building the world's first nuclear reactor: Chicago Pile-1.

The success of Chicago Pile-1 proves it is possible to get steady energy out of reactors, and countries rush to build their own. Nuclear electricity becomes ridiculously cheap and quickly pops up all over, from Brazil to the United States to India. In a world with unlimited access to clean and cheap energy, things that sounded like science fiction become everyday reality.

Within twenty years of the discovery of nuclear fission, London becomes completely different from the smog-filled mess it once was. The city is so clean you could eat off the streets. Well, not literally, but you know what I mean. As countries use less and less fossil fuel, deaths caused by air pollution dwindle everywhere. Lifespan goes up and things like asthma enter the category of "weird conditions people used to have in the past," like tuberculosis.

In the United States, New York City becomes the poster child of this new era, featuring huge skyscrapers, some powered by mini reactors, sparkling against the night sky. Balconies overflow with plants, creating a sort of vertical jungle. The Hudson River, once a toxic waste dump, is now so clear you could practically see the fish high-fiving each other. Central Park, blessed with fresh air, lush greenery, and the sound of birds, becomes the go-to spot for people needing a break from the city's chaos.

Transportation has a makeover too with electric vehicles kicking gas-guzzlers to the curb. High-speed trains zoom along magnetic tracks, connecting cities and countries, making travel incredibly easy. The annoying sounds of cars, motorcycles, and leaf blowers are replaced by the sounds of nature and people talking to one another. Nuclear-powered ships crisscross the oceans, delivering goods everywhere without spewing out pollution.

Because nuclear power plants take up such a tiny amount of land, green areas spring up around cities and forests flourish, while the planet begins to heal from the gross effects of the Industrial Revolution. Nuclear electricity becomes the backbone for incredible new clean energy tech. It allows us to come up with scaled-down wind turbines that seamlessly blend into buildings. Solar cells cover roofs and surfaces, collecting energy from the sun. Cities become truly clean and green as coal, oil, and methane gas become ancient history.

Economies boom as they have a surplus of energy and don't have to import fossil fuels, allowing nations to put a lot more money into education, healthcare, and research instead. Global cooperation goes up as countries swap nuclear electricity tech and know-how, creating an atmosphere of unity and progress. Energy independence also cuts down the clout of oil-rich nations and leads to a more balanced world stage. Energy conflicts fizzle out, replaced by teamwork on nuclear safety and waste management.

Within forty years of the discovery of nuclear fission, every single human on Earth has access to electricity. Okay, not everyone. There is a small group of people who choose to live "the old school way" by using only firewood and not embracing electricity. But hey, to each their own.

Going all in on nuclear eventually enables us to unleash the full potential of artificial intelligence to help solve the world's hardest problems. We find the cure to most cancers, eliminate all genetic diseases, and keep making people healthier. Nuclear also gives us the electricity needed to recycle even the most difficult things like electronics, plastics, and batteries. With rockets powered by nuclear thermal propulsion, we can explore the universe faster and settle on other planets. Tiny nuclear reactors power human colonies on the moon and Mars, making humans an interplanetary species.

Nuclear fission completely transforms the course of human history. With a cheap, plentiful, and safe source of energy accessible to all, humans enter an unprecedented chapter of prosperity, abundance, and peace.

WHAT ACTUALLY HAPPENED

You might be scratching your head right now, wondering if I've taken some hardcore drugs or if we live in different universes. Our planet could have looked closer to what I just described in italics, had the discovery of nuclear fission happened in a different year or a different place. But, tragically for us, Hahn and Strassmann's night at the lab happened during one of the worst times ever. You don't need to be a history buff to know that 1938 in Germany was a "bad place, bad time" situation—a time when a dictator was working to build up his forces and weapons in preparation for his megalomaniacal ambitions.

As the two chemists studied in their lab, Nazis marched on the streets outside.

Lise Meitner was one of their closest colleagues. The only reason she wasn't in the lab with them that night was because she had fled to Sweden to escape persecution, and possibly death, for being Jewish. World War II would start one year later.

While physicists around the world were buzzing with excitement over the discovery of nuclear fission, some of them also saw the writing on the wall. Controlled splitting of atoms could provide useful energy to power people's lives. But uncontrolled splitting of atoms could create a weapon of mass destruction unlike anything the world had ever seen.

In August 1939, less than a year after the discovery, celebrity physicist Albert Einstein cowrote a letter with physicist Leo Szilard to American president Franklin D. Roosevelt sounding the alarm about the potential dangers of this new breakthrough. He warned that Germany was most likely already trying to develop a bomb using the underlying science. Imagine the level of destruction and suffering Hitler could have inflicted upon the world, had he gotten his murderous hands on nuclear weapons.

Eventually the rumors motivated Roosevelt to kick-start a massive effort to beat Nazi Germany to the nuclear bomb punch. The Manhattan Project launched in 1942 under the leadership of J. Robert Oppenheimer. Lise Meitner was invited to join the effort but refused, saying, "I will have nothing to do with a bomb!"

The effort culminated in the bombings of Hiroshima and Nagasaki in 1945, which effectively ended World War II. It's estimated that between 129,000 and 226,000 people, mostly civilians, were killed. That would be the first and last time, so far, that nuclear weapons have been used in battle.

Just *two* years after the chaos of World War II, the planet's biggest superpowers decided to keep the drama going by getting straight into the Cold War. Because apparently, peace and harmony were just overrated. During this time, the United States and Russia engaged in an arms race, stockpiling a bunch of nuclear weapons and creating a whole new anxiety genre: nuclear annihilation, the fear that a weapons exchange would wipe humans off Earth. The national security concept was MAD—Mutual Assured Destruction—meaning that any country that used nuclear weapons against its enemy would be wiped out as well.

For obvious reasons, this fear scarred a whole generation, especially those who had to do "duck and cover" drills. Picture American kids in the 1950s and 1960s in school doing their thing. Suddenly, an alarm goes off and the teacher yells, "Duck and cover!" They drop everything and lunge under desks, hands over their heads. The geopolitical tensions were so high at the time, everyone feared and prepared for a nuclear weapons attack at any minute.

It didn't seem like an overreaction back then. The photos of the aftermath of the bombings in Japan had become seared into public consciousness and the wars in Korea (1950–53) and Vietnam (1954–75) kept everyone on edge.

To say that the timing of the discovery of nuclear fission was shitty would be an understatement. It *could* have changed everything for the better. But the world was introduced to nuclear technology through horrific images of mushroom clouds and terrified children crying, running away from crumbling buildings. Besides the awful devastation, the dropping of the bombs resulted in unexpected collateral damage, which ended up causing even *more* harm for humans.

THE IMPACT

The public could not disentangle images of mushroom clouds and frightening duck and cover drills from the peaceful use of nuclear to generate electricity. And how could they? Governments made everything worse, as they cast a veil of secrecy over all nuclear research. From the 1940s until 1954, only the military could build and operate reactors. As a result, anything with the word "nuclear" in it became something to fear, and we ended up villainizing one of the best sources of energy we have.

The full story of how we got to be so anti-nuclear electricity is a lot more complicated—and wildly fascinating. It's filled with mystery, unexpected events, and celebrities. I'll dive into all the juicy details in chapter 7, but for now all you need to know is this: In large part because of the bad rep it got during the Cold War, nuclear electricity hasn't had the opportunity to reach its full positive potential.

Instead, in the second half of the twentieth century, the world doubled down on fossil fuels as our primary energy source, developing an addiction that we still haven't been able to kick and that's costing us immensely. In a single year the pollution from burning these fuels causes at least four million deaths, which is way more than nuclear electricity has ever caused in its entire *history*.

Our addiction to coal, oil, and methane gas has polluted communities, claimed millions of lives, and set the stage for one of the biggest threats our civilization faces: climate change. At the same time, energy inequality and poverty remain serious issues.

Is it surprising that so many people, especially young people, feel anxiety and dread about the future? They're living in a world that seems to be inching closer and closer to total environmental and societal collapse.

INFINITE POSSIBILITY

Energy poverty is a topic that gets me fired up. That's because I was born in 1990 in Brazil, in a small town about ninety miles north of the border with Uruguay. Though my family was middle-class, my apartment had few modern amenities because electricity was very expensive and sometimes unreliable. To give you an idea, in the 1990s the United States created eight times more electricity per person than Brazil. We had all the basics—lights, a gas stove, a refrigerator, and a standing fan. But other tech that people in North America take for granted, like air-conditioning, dishwashers, and laundry machines, was something only the wealthy could afford.

As I got older and electricity got cheaper, we started adopting more and more of these things. But it was only when I moved to the United States at age eighteen that I truly understood what a high energy life looks like. Two years prior to this move, my life had been turned upside down. In case you are like me and don't read book introductions, I will tell this story again. When leaving my high school, I was approached by a man who invited me to participate in a contest for one of Brazil's biggest modeling agencies. Long story short: I placed third in the nation and began working as a fashion model. As my career took off, I started traveling the world and ended up in sunny Miami. I'll never forget stepping into the huge building that would be my home for years to come. All the residences had air-conditioning, a dishwasher, a laundry machine, and a dryer. In hindsight, even though it was a very mid apartment, it felt like a mansion.

My unique experiences gave me an understanding of how life changes as people move through different levels of energy access. Let me break it to you: Life is a lot more fun when you don't have to spend hours hand-washing clothes.

By my mid-twenties, I was traveling the world, living many women's dreams of being a fashion model. I was also building a promising beauty company. But under the surface, a sense of doom was building. The belief that humans were awful, a cancer on this planet, permeated my worldview. Like so many, I believed we had destroyed Earth and would never be able to get out of this mess. Everything felt hopeless. I didn't even want to have children—why bring them into such a horrible world?

As I mentioned in the introduction, my outlook completely shifted when I read *The Beginning of Infinity* by British physicist David Deutsch. He makes a radical case for *pragmatic* optimism that resonated with me and has since changed my idea of what's *possible*. I'm going to try to sum it up.

Deutsch believes humans are special. Not because of supernatural properties, but because we're the only species—we know of—that can learn about the world and use that knowledge to change it. In an interview with Naval Ravikant, he explained that "humans have explanatory creativity. Once you have that, you can get to the moon. You can cause asteroids which are heading toward the earth to turn around and go away. Perhaps no other planet in the universe has that power, and it has it only because of the presence of explanatory creativity on it."[1]

Think about how people used to die from infections and diseases simply because we didn't know about germs. It wasn't until the late 1800s that someone finally figured out tiny organisms were responsible for spreading disease. Once we had that knowledge, we came up with easy solutions—like handwashing and antibiotics—for a problem that had baffled our ancestors for thousands of years.

Deutsch argues this process opens a door to *infinity*. In theory, there are no limits to the amount of knowledge we can create, which means every problem can be solved if we have enough information about it. In our endless creativity, we are only constrained by the laws of physics. To me, this is the most optimistic view of the world, which I like to summarize as: Anything (that doesn't violate the laws of physics) is possible.

But perhaps the most powerful idea in his book is that utopia has never and will never exist. There hasn't been a single time in history when humans lived in complete harmony, either among themselves or with the environment. Our species had to endure ice ages, droughts, predators, starvation, wars. Essentially, our existence is marked by solving challenges that seem to get more and more complicated as we develop more sophisticated tech. The best we can do is to make incremental progress and hopefully leave our descendants with at least a different set of issues than our own.

Adopting this framework meant that instead of feeling down about problems, I began seeing them as *inevitable*, but ultimately *solvable*. That's why in 2019, when I decided to do something to help tackle climate change, I didn't want to focus on the issue itself, but instead on its solutions. As I learned about them, one caught my attention: a source of energy that can create *massive* amounts of clean, reliable, always-on electricity—all while leaving plenty of space to plant more trees, farm food, and let nature heal. It creates loads of high-paying stable jobs that boost local economies. Despite its bad reputation, it's also incredibly safe and has already proven that it can slash fossil fuel use. Though we missed the chance to unleash its full potential in the past, it can still help us build a rad future.

Of course, I am talking about nuclear electricity.

TL;DR

Ignore the human-hating nonsense that's pervasive in our culture. Our species' unique ability to acquire knowledge, change the entire universe, and solve problems makes us one of the coolest things out there. Climate change is a pretty hard problem, but 100 percent solvable. We have all the tech we need, and one really stands out: nuclear electricity. Unfortunately, the nuclear age was kicked off by two scientists in Nazi Germany, so instead of clean energy for all, we got nuclear bombs and a world where kids had to duck and cover under school desks, which frankly doesn't seem like a great survival strategy against atomic blasts. Thanks in part to this cosmic mix-up, we've gotten even more hopelessly addicted to burning fossil fuels to power our civilization—an addiction that's costing us millions of lives every year and causing climate change. But the promise of the atom isn't dead.

PART 2
ALCHEMY

2: HUMANS AND ENERGY

ENERGY IS THE CURRENCY OF THE UNIVERSE

Nuclear is the best source of energy. But before we dive into why it's the coolest and most important technology of our lifetimes, you'll need to understand what *energy* is and why we need it.

If you're like most well-adjusted people, you probably haven't spent a lot of time thinking about this topic. That's normal, we don't spend much time thinking about the air in our lungs either. But just imagine waking up one day to discover that aliens had zapped all energy away. Lights out. Cars dead. No Netflix. No more delivery apps. Sure, the novelty might be fun for a day or two, but soon enough terror would set in. How would you work? Or feed your family? How would food even get to stores without trucks, ships, and planes?

If you look the word "energy" up in the dictionary (don't worry, I did it for you), you'll see a snooze-fest definition about "a fundamental entity of nature that is transferred between parts of a system in the production of physical change within the system and usually regarded as the capacity for doing work."[1] No offense to Merriam-Webster, but I'll stick with the definition used by writer Tim Urban: Energy is "the thing that lets something do stuff."[2] Without it, not a whole lot would be happening.

All of the energy in the universe—seriously, all of it—came from the big bang, that mysterious explosion that gave birth to space and time as we know it some 13.8 billion years ago. The universe is still packed with that initial explosion of energy—every single bit of it is still out there. This energy does stuff like helping stars shine or holding planets together, but something really special happens on Earth. It is the only place that we know of where that cosmic energy also powers *life*. Plants, insects, reptiles, mammals, and more all evolved with basically one goal: Get as much energy as possible and reinvest it into reproduction.

But most creatures on our planet can't make their own energy, a problem they solved by becoming thieves. It goes like this: Almost all energy on Earth comes from the sun. Plants soak up that

sunlight and turn it into plant food. Herbivores eat the plants and steal *their* energy. Then carnivores eat herbivores to steal *their* energy. And so on. Every creature is snatching energy from another creature, in a chain that connects all the way back to the sun.

Now, not to brag, but humans are quite exceptional when it comes to using energy. For hundreds of thousands of years, our ancestors were just minding their own business, stealing juice from plants and other animals to get the energy they needed to do basic stuff like running, moving rocks around, and making babies. But like every other mammal, they were limited to using the energy they had stored in their own bodies.

Everything changed when our ancestors realized they could harness the energy *around* them too. The first source they tapped into was fire. We don't know exactly when this happened, but evidence suggests they were regularly playing with it around half a million years ago. Fire allowed them to cook food, a little hack scientists believe propelled early hominid brains into *Homo sapiens* territory. Cooked food is easier for our bodies to break down, meaning less energy is wasted on digestion. Having those extra free calories allowed us to develop giant, juicy brains. Fire also provided warmth, comfort, and eventually the ability to melt and shape metals to make better tools.

But fire was just the beginning. It kicked off a bunch of shortcuts our species has used to gain the ultimate prize in life: *time*. Getting more calories from cooked food meant we could spend fewer hours foraging and hunting to survive. Humans suddenly found themselves with extra brainpower and free time to imagine better ways of living. Tools and gadgets soon followed. From waterwheels and pulleys in the olden days to the cars and laundry machines we have today, humanity escaped survival mode. We had hours, days, and even years freed up to focus on much more fun things. With all that time, humans started doing what we do best: making art, building cool stuff, exploring the world, and dreaming of the stars. Okay, and maybe creating more chaos too.

But here's the thing: None of these time-saving inventions work without a constant stream of *energy*. They can't magically make it themselves, or steal it from other life-forms, like we can. This has led us to create entire industries devoted to harvesting, harnessing, and channeling energy to feed our ever-hungrier technologies.

ENERGY TIMELINE

Back in the day, our machines relied on pure mechanical power. Think a bunch of sweaty peasants pulling levers, mules dragging heavy carts, water turning wheels, and so on.* Then we got into burning biomass, which is just a fancy word for wood, plant leftovers, and even animal poop. But after the Industrial Revolution in

---

* Here's a fun fact about those mechanical power days (or an extremely not-fun fact, depending on whom you ask): In medieval kitchens, dogs were bred specifically to run on giant hamster wheels used to power rotating spits that roasted meat over the fire. They were called "turnspit dogs."

the late eighteenth century, things got crazy. This was a time when manufacturing shifted from manual to machine-based production. Suddenly, steam engines and textile factories were all the rage, and they needed obscene amounts of energy to keep going. That's when we turned to the dark side: fossil fuels. Coal, oil, and methane gas became the life force behind our inventions. Fast forward to 2024, and almost all our machines still guzzle them, from cooking stoves to cars and ships. Over 75 percent of the energy used in the world is generated from fossil fuels.³

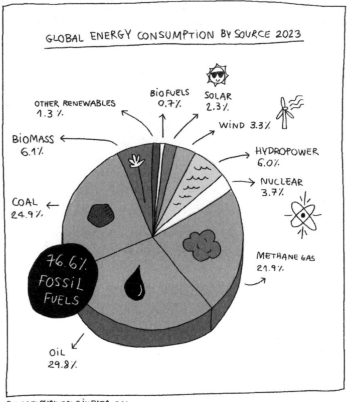

SOURCE: OURWORLDINDATA.ORG

Honestly, fossil fuels would be epic if they weren't killing us. They're made of dead plants and animals that lived hundreds of millions of years ago. (I'm sure the dinosaurs would be stoked to know how many lawn mowers they've powered.) When defunct plants and animals are buried deep underground and subjected to heat and pressure over millions of years, their molecules start to break down. This process eventually creates coal, oil, or methane gas. When we burn these fuels, they release ancient stored energy.

You might be surprised to learn that coal was once a win for the environment. In the sixteenth and seventeenth centuries, wood was being used to make charcoal for smelting iron (and other things) so rapidly that European forests were being destroyed. Using coal back then meant more trees stayed standing.* Fortunately for human civilization, the discovery of coal as a powerful energy source fueled the Industrial Revolution. Unfortunately for, well, human civilization, coal and other fossil fuels *also* set the stage for most of the modern environmental challenges we face today.

The widespread use of coal and, later, oil and methane gas fueled mind-blowing levels of technological, economic, and social progress. But the same fossil fuels that enabled our civilization have been making us sick and poisoning the planet for centuries. Burning them spews out nasty stuff like sulfur dioxide, lead, mercury, and other tiny pollution particles into the air. Breathing in this pollution can wreck your health, causing asthma, chronic obstructive pulmonary disease (COPD), and heart issues. More than four million people die prematurely every year because of pollution-

---

* You need about 10 pounds of wood to get around a pound of charcoal, and the conversion process gives you less than half the energy you started out with.

related health issues. A study found that breathing polluted air regularly is just as bad for you as smoking cigarettes.[4]

On top of creating pollution that's bad for our health, burning fossil fuels releases greenhouse gases like carbon dioxide and methane. These gases trap heat in the atmosphere and mess with our planet's natural climate patterns. As a result, sea levels are rising, threatening coastal communities and island nations. Weather patterns are becoming more erratic, leading to more frequent and severe hurricanes, droughts, and floods. Biodiversity is under siege as species struggle to adapt to the changing climate. The very fabric of our planet is being reshaped by our toxic attachment to fossil fuels.

So, thanks for all the good times, oil, methane gas, and coal, but it's time to go bye-bye. Getting rid of these fuels is a must, but it won't be a small feat, as they are still deeply intertwined in our modern lives.

There's hope though, because in the late 1800s we tapped into a very special source of energy: electricity. It was efficient, easy to transport, clean, and didn't smell like prehistoric barbecue. We started using it to illuminate our streets and homes with light bulbs, to power factories, and eventually to communicate through inventions like the telegraph and telephone. Electricity today keeps the lights on in our homes, charges our electronics, and even powers some cars. But most of our trucks, ships, cars, and heaters still run on fossil fuels, which is why you might have heard about "electrification." We want to power more stuff with electricity and less stuff with toxic sludge.

While electricity comes into our homes as a clean and efficient power source, it often hides a dirty little secret. Even though an electric car might not be spewing pollution into the streets as it cruises by, it could very well be getting its juice from a coal power plant. That's because electricity is a *secondary* source of energy, which means it must be created using a *primary* source of energy. These primary sources include fossil fuels like coal, oil, and methane gas, as well as alternative sources like solar, nuclear, wind, geothermal, and hydropower.[5]

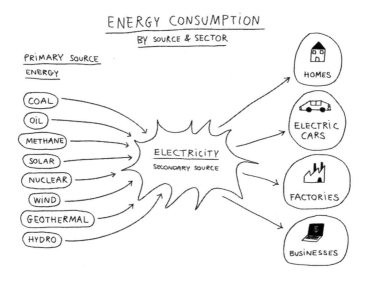

In 2023, more than 60 percent of the world's electricity was created from fossil fuels. That percentage was the same in the United States specifically.[6]

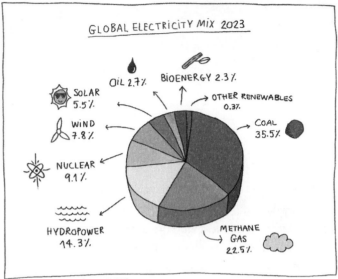

SOURCE: OUR WORLD IN DATA.ORG

So how do we break up with fossil fuels for good? It's a two-step dance. Step one is to electrify everything that currently uses oil, methane gas, or coal. Your car? Electric. Your stove? Electric. Your heater? You guessed it—electric. You get the idea; if it currently runs on fossil fuels, we want to make it run on electricity instead. Step two is to generate all our electricity using *clean* primary energy sources only—ones that cause less harm to the planet and people.

This is where nuclear electricity comes in. Sure, it had a bit of a rough start in the twentieth century, but it's time to give it a second chance. Let's wipe the slate clean and take a fresh look at this misunderstood queen.

TL;DR

Energy is the currency of the universe. Not in like an annoying woo-woo way. We need it to be able to do anything. In our quest for more and more energy, we've tried pretty much everything out there: from waterwheels to burning wood, poop, and even liquefied ancient fossils. While burning this stuff has given us unimaginable levels of technological development and wealth, the truth is that we can do better. Humans have outgrown fossil fuels. But like any codependent relationship, it's not that easy to walk away. Enter electricity. Electricity is clean and efficient, but it's got a dark little secret: Most of it is made from burning coal, methane gas, or oil. The solution is to electrify everything and source all that power from clean options like nuclear. Yes, nuclear—the misunderstood energy source that's ready for its comeback.

3: FOSSIL FUELS' FINAL BOSS

To move away from fossil fuels, we are going to need a lot of *clean* energy, and nuclear will play a big part. How does it work, and why is it considered clean? This next section is about to get science-y, but I promise to make it fun.

Remember that illustration of an atom from science class? The one with a ball in the center and some rings around it? Each ring represents the orbit of an electron, making it a bit like our solar system in miniature. The nucleus is like the sun, and the electrons are like the planets orbiting around it.* Nuclear energy is the energy that's trapped inside the nucleus of atoms, like a treasure chest waiting to be unlocked.

There are two ways to get this energy out. The first is by smashing *small* atoms together, like stacking smaller LEGO pieces to

---

* IRL, those electrons don't actually move as predictably as moons or satellites. They form more of a fuzzy cloud surrounding the nucleus, where we can only guess the location of any given electron at any given time.

make a bigger one. This is called "fusion," and it's what happens in the sun. The second is by splitting big atoms into smaller pieces, like breaking a big LEGO piece into smaller ones. This is called "fission" and it's what Fritz Strassmann and Otto Hahn were unknowingly doing in their lab back in 1938. While a bunch of smart people are trying to build fusion reactors, they're still far from becoming everyday reality.* In 2024, fission reactors are still the only type of reactor producing electricity worldwide. That's why I will focus on nuclear fission in this book.

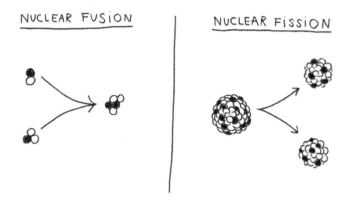

When we split an atom's nucleus, energy in the form of heat comes out, and we can use that heat to make clean electricity that goes on to power our lives.

---

* Before fusion fans get pissed off, let me clarify that I am super excited about fusion and think it will be an important source of energy in the future. But this book is focused on tech that can be scaled right now.

HOW NUCLEAR REACTORS WORK

Picture a kettle on a stove: The water heats up and steam comes out. That's essentially how all power plants work—they're just super-sized kettles that can make electricity. They heat water, create steam, and use it to spin fanlike turbines that are hooked up to electric generators. Nuclear power plants do the same thing. But they are unique in that instead of using the heat from burning fossil fuels to boil water, they use the heat from splitting atoms. It's like if your kettle was heated by magic spicy rocks instead of an open flame.

And here is where nuclear electricity shines: Splitting atoms doesn't release the pollutants that are let out when burning fossil fuels—nasty stuff like lead, mercury, and sulfur oxides or planet-warming gases.

That's why *nuclear electricity is clean*. It's like having your cake,

eating it, and it being good for you too—you get the power you need without the guilt.

Most existing nuclear reactors are what are called "light water reactors," which means they use water for sustaining a chain reaction and for cooling. That's the design I'll focus on in this book, but just know that there are seven different types of nuclear power reactors around the world.*

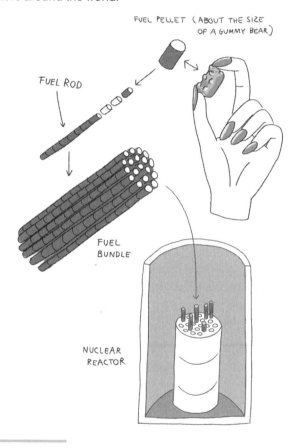

* These are reactors whose main job is making electricity. But there are also research reactors, which are used for, well, doing scientific research education and for producing isotopes that can be used in medicine, like for cancer treatments.

The fuel used in reactors is made of a special kind of uranium that's really good at splitting apart. This uranium fuel comes in tiny pellets, each about the size of a gummy bear. But instead of being squishy, it's a hard ceramic. These pellets are stacked in thirteen-foot-long hollow metal tubes that look like giant straws. They're bundled together and put in the reactor vessel, which is like a big pot filled with water.

If you want to learn more about radiation and uranium, the fuel used in nuclear reactors, I'll explain it in the nerd alert below. But no pressure, you can still understand nuclear electricity without it. Follow your bliss.

NERD ALERT!

> Ever wondered what's inside an atom's nucleus? It's like a tiny party with protons and neutrons. Protons are the life of the party, always positive and charged up. Neutrons, on the other hand, are neither positive nor negative, they're just vibing. Since protons have the same positive charge, they naturally want to push one another apart, like magnets with the same pole facing each other. What keeps them together is the strong nuclear force, one of nature's fundamental forces, while neutrons help stabilize the nucleus by adding to this force. Most atoms we see in nature are stable, meaning their nuclei maintain the same number of protons and neutrons over time. But in the late 1800s, we discovered that some atoms have unstable nuclei and transform into different, more stable atoms.

Don't worry about why this happens. Let's just say that it's because of complicated, physics-y reasons.*

Like most of us, the nucleus of an atom really *wants* to be stable. If things aren't in balance, it'll start getting rid of its extra bits like someone doing a Marie Kondo closet clean-out. That's what radiation is—just a floofy atomic nucleus getting rid of unnecessary stuff in an effort to achieve a balanced life.

There are around three thousand known unstable isotopes, also known as "radioisotopes." Imagine isotopes as different flavors of ice cream, where each flavor has its own unique taste, but they're all still ice cream. Just like how each ice-cream flavor has a different taste due to the added ingredients, isotopes of the same element have different atomic masses because of the varying number of neutrons in their nuclei. However, they still share the same number of protons and the same chemical properties, so they're still the same element.

A radioactive element is an element that has no stable isotopes—all possible variations of that element are unstable and radioactive. Several dozen of these elements occur in nature, with uranium and thorium being among the most abundant. Uranium has three natural isotopes (uranium-238, uranium-235, and uranium-234) and it's used as the primary fuel in most nuclear power plants.[1]

---

* This is how a physicist told me I should explain this.

The core of a nuclear reactor is where the real magic happens. Tiny particles called "neutrons" collide with uranium atoms in the fuel. Much like a triggering text that sends you spiraling, these collisions push the already unstable uranium atoms over the edge, causing them to split apart. That splitting releases a ton of heat and more neutrons that barrel straight into neighboring atoms, splitting those too. It's like atomic dominoes, with each breakup triggering more chaos, creating what we call a "chain reaction."

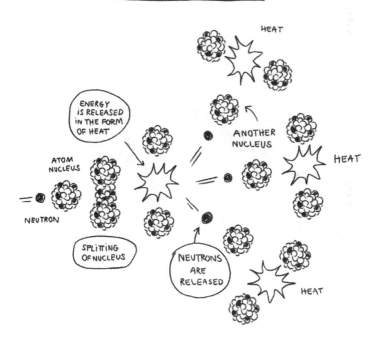

A nuclear reactor's main job is to kick-start and control chain reactions, keeping the amount of heat and steam levels in check. To keep running smoothly, reactors come with some accessories.

The core lives inside some kind of containment, usually a dome-shaped structure made of reinforced concrete, steel, or lead.* This little fortress protects the reactor from external threats and minimizes risks of contaminating the outside world in the very rare event of an accident.

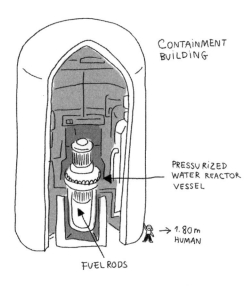

Because of the huge amount of energy that nuclear fission generates, reactors get extraordinarily hot and need to be cooled down to keep working properly. Almost all reactors use water for that. That's why most nuclear power plants sit next to rivers, lakes, or oceans. They pull in water, use it for cooling, and return it right back to the source—just a little warmer than when it went in.

Inside a nuclear reactor, there are two separate water systems

---

* Some reactors, like the one in the infamous Chernobyl power plant, don't have a containment structure. Which is just... why?

working together. The water inside the core gets heated as it flows past the fuel rods. That heat is transferred to a completely separate set of pipes carrying fresh water. Think of it like a relay race: The reactor water "passes the baton" of heat to the fresh water without the two ever mixing.

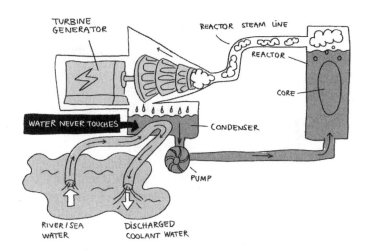

Some nuclear power plants also feature cooling towers. These are the big concrete structures releasing water vapor that you've probably seen when driving by a power plant (or in *The Simpsons*). In this case, instead of sending the heated water back to its original source, the plant pumps it to the cooling towers. There, the water cools off and gets ready to be used again.

A lot of people assume cooling towers are spewing radioactive smoke up in the air, but that's not true. The cooling water never touches anything radioactive, so what you're seeing is just clean water vapor. Cooling towers are basically cloud-making machines.

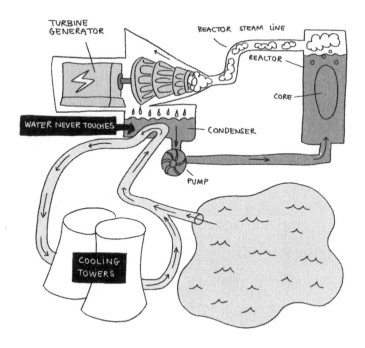

As of 2024, there are around 440 nuclear reactors worldwide,[2] generating enough juice to power the equivalent of three and a half Brazils. Nuclear is the second-largest source of clean energy globally, just after hydropower. In the United States, nuclear plants provide about 20 percent of the nation's electricity, making up nearly half of all its *clean* sources.

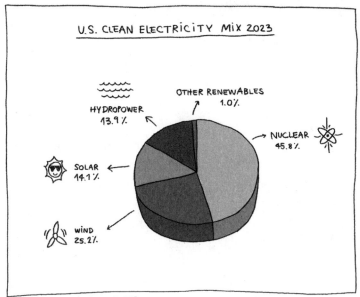

SOURCE: OUR WORLD IN DATA.ORG

NUCLEAR IS RAD(IANT)

Now that you know nuclear reactors are just high-tech machines for creating pollution-free heat, steam, and electricity, let's dive into why nuclear is arguably the best source of clean energy we have.

IT'S ALWAYS ON

The sun isn't always shining, and the wind isn't always blowing. Nuclear is unbothered by the weather. It keeps chugging along, day and night, rain or shine, making a steady stream of electricity whenever we need it. Because of that, it has the highest capacity factor of any source of clean energy. Here's a simple way to understand capacity factor: Imagine you own a donut shop. The maximum number of donuts your shop can make in a day is one hundred. But because of a shortage of flour, a couple of workers being sick, and some machines needing maintenance, you end up only being able to make twenty. That would make your donut shop's capacity factor 20 percent for that day.

In 2023, American nuclear reactors had an average capacity factor of over 93 percent, which means they were making the maximum amount of electricity they can make over 93 percent of the time.[3] But this number varies depending on the country, the reactor's design, and the fuel. For example, France ramps its nuclear plants up or down depending on the amount of electricity from renewables on the electric grid. If solar panels and wind turbines are producing a lot of electricity, they ramp nuclear down. For this reason, France's capacity factor is around 77 percent.

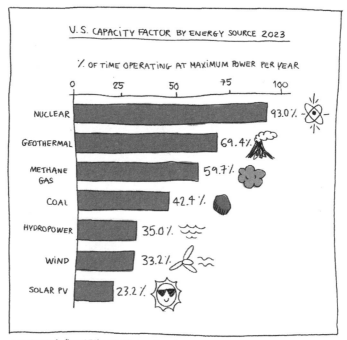

SOURCE: EIA, "ELECTRIC POWER MONTHLY"

We've run nuclear reactors deep underwater in submarines, in the remote cold and dark of Antarctic winters, in the desert, and even in space.[4] Nuclear is reliable and adaptable to pretty much any setting or scenario you can imagine.

That's why the U.S. Department of Energy (DOE) has established that to get to net zero by 2050,* the country must triple the number of reactors it has.† The DOE is stoked about nuclear's role as a "firm" source of clean energy, which is just a weird way of saying it's dependable and always has your back.[5] In a 2021 study, researchers found that in California, solar and wind energy supplies can drop by up to 60 percent between summer and winter months because of cloudier skies and less-powerful winds.[6] While batteries can help on a bad day, they can't pull the state through several days or weeks of low supply from solar panels and wind turbines. The study concluded that to get to a 100-percent-clean electric grid quickly and cheaply, California would need around 30 gigawatts of "firm clean power"—the equivalent of thirty big nuclear reactors.

IT DOESN'T TAKE UP A LOT OF SPACE

Nuclear's biggest flex is that it packs a huge punch in a tiny space. Picture the fuel used in most reactors: a baby ceramic pellet, as small as a gummy bear. But don't let its smallness fool you—nuclear is *one million times* more energy dense than fossil fuels like coal or oil. In fact, each pellet has the energy equivalent of 149 gallons of oil or 2,000 pounds of coal.[7] Using just one of them, you

---

* Net zero is a global goal to make sure we don't pump more greenhouse gases into the air than we can remove. It's about both emitting less *and* removing what we do release (with things like carbon capture machines) to reach a net balance of zero emissions.

† Lucky for us, the DOE has concluded that we can hit the target of 300 gigawatts in the nick of time—if we bother to try.

could keep the lights on in an average American home for about 2.5 months. That's 2.5 months of bingeing TV series, scrolling on Instagram and TikTok, all powered by a tiny piece of uranium.[8]

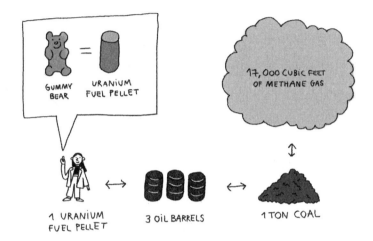

This fuel is so dense, it's almost impossible to wrap our minds around. That's why nuclear isn't just efficient—it's the most land-efficient energy source we've got, period. The global data powerhouse Our World in Data crunched the numbers on the land footprint requirements of different energy sources. These calculations even included the amount of land used for mining. Turns out that to crank out the same amount of electricity, nuclear needs fifty times less land than coal. But it doesn't stop there. Even when you stack nuclear against the darlings of the clean energy club, it still leaves them in the dust. Solar gobbles up between eighteen to twenty-seven times more land to make the same amount of electricity.*

---

* This number is for solar PV installed on the ground.

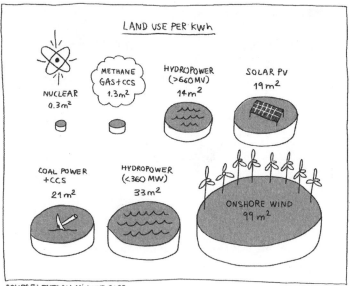

SOURCE: ENERGY MINUTE 2023

In 2023, the American environmental group Clean Air Task Force dropped a report on what California would need in order to speed up its clean energy rollout. It concluded that to reach a zero-carbon electricity grid using solar alone, they'd have to cover an area the size of Los Angeles and San Diego *combined* with solar panels. While the state might technically have that much open land, the report warned it would be a logistical nightmare unless they're cool with taking over natural habitats or steamrolling marginalized communities.[9]

Scientists from MIT and Stanford also found that to replace Diablo Canyon, California's only nuclear power plant, with solar panels, it would take ninety thousand acres of land. To put things in perspective, that's more than six hundred times the amount of land the plant takes up.[10]

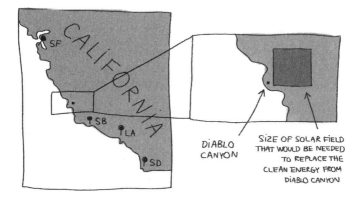

If covering massive swaths of the planet with panels and turbines were the *only* way to ditch fossil fuels, I'd be all for it. But thankfully, we've got options—a whole buffet of clean energy choices to pick from. Let's put solar panels on roofs and wind turbines in deserted lands. Nuclear plants are a vibe because they sit compactly, leaving a lot more room for nature. Love that for us.

IT'S HOT

While most clean energy sources are basic (they can only make electricity), nuclear is unique. Reactors can make both electricity and heat—like, industrial-grade heat. The kind of heat you need to make paper, crank out fertilizers, or manufacture a bunch of the stuff modern life depends on. While in 2024 most of this demand was met by burning fossil fuels, nuclear can change that. For example, heat from reactors can be used to produce hydrogen for fertilizers without releasing greenhouse gases. We can also use it to desalinate water, which is a must for dealing with climate change–induced droughts and hydrating a growing population. It can even directly warm up homes and buildings through something called "district heating." Forty-three reactors around the world already do that on the daily.

Why not give a little love to an energy source that can multi-task?

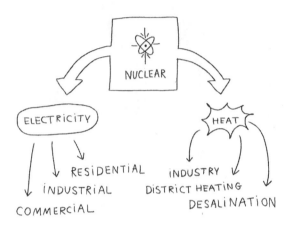

IT HAS A PROVEN RECORD OF SLASHING FOSSIL FUELS

One of the reasons nuclear is such a no-brainer is because we know it can replace fossil fuels—it's already done it before. Just look at Ontario, Canada: Back in 2005, coal made up a quarter of the province's electricity. But with most of their hydropower maxed out, they couldn't just swap coal plants for dams. So, they decided to modernize three existing nuclear reactors and get them back online. That clean power allowed Ontario to shut down all its coal plants by 2014. By 2019, roughly 92 percent of their electricity was clean.[11]

This phaseout remains the biggest greenhouse gas reduction effort on the American continent.

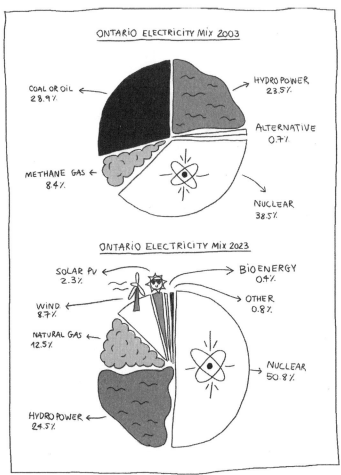

SOURCE: ONTARIO ENERGY BOARD

We can't talk about a serious energy glow-up without giving a nod to the ultimate nuclear queen: France. In the 1970s, facing an oil crisis and a heavy reliance on fossil fuels, France went all in on nuclear and kick-started one of the most successful clean energy transformations the world has ever seen.

By the end of this build-out, fifty-nine reactors were churning out over 70 percent of the country's electricity. France's CO_2 emissions (per unit of electricity produced) went down by a jaw-dropping 79 percent in a decade. While others are busy talking about cutting emissions, France has already shown us how it's done.

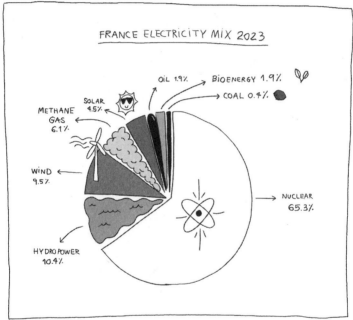

IT'S CLEAN
Reactors don't spew greenhouse gases or toss particulate matter into the air. But some people raise questions about emissions from activities like mining fuel or building nuclear power plants.

That's because it takes energy to dig up raw materials, build infrastructure, keep things running, and eventually decommission it all. Most of the machinery doing that work—digging, hauling, building—is still powered by fossil fuels. But that's true of *every energy source*, even the clean ones.

Here's where nuclear stands out: It needs far less of everything to get the job done. This boils down again to its incredible energy density, which means only a tiny amount of fuel is needed to produce a unit of electricity. Plus, nuclear plants can run for up to eighty years, reducing the need for frequent rebuilding and the materials that come with it.[12]

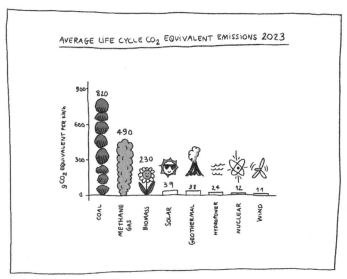

SOURCE: IPCC

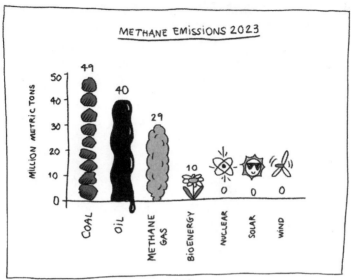

SOURCE: IEA, GLOBAL METHANE TRACKER

IT LETS US REUSE OLD FOSSIL FUEL FACILITIES

Almost a quarter of the coal plants in the United States are supposed to be retired by 2030. Good riddance. The great news is many of them could be converted into nuclear power plants. In 2024, the Department of Energy studied four hundred of these soon-to-be-shut-down coal plants and found that approximately 80 percent of them had the potential to be repurposed as nuclear plants.[13] Talk about a win-win. Globally, over three thousand coal plants could potentially make the same switch.

This repurposing is good for both the environment and the economy. Converting a coal plant into a nuclear facility could cut build costs by up to 35 percent compared to starting from scratch. Even if equipment can't be reused, simply using existing power lines, roads, and infrastructure can still shave off about 10 percent of costs.[14]

And there's more. One of the biggest obstacles to clean energy rollout is the need for new high-voltage transmission lines, which is a major headache. Nuclear can sidestep this issue in two ways: First, reactors can be built close to where the energy is needed, unlike wind or solar, which often need to bring in power from far-off places, adding transmission needs. Second, nuclear's high power density means existing transmission lines can carry more energy per mile, making the grid more efficient without all the extra lines.

IT GUARANTEES A JUST TRANSITION
Turning old coal plants into nuclear facilities won't just be cheaper—it's a win for jobs and the local economy. Sure, you could decommission a coal plant and toss up some solar panels on the land, but that'll only replace a tiny fraction of the power and offer next to nothing for the community. That's because solar or wind farms don't need a lot of employees. Repurposing coal sites as nuclear plants, on the other hand, gives marginalized coal communities a real shot in the energy transition, with long-term, high-paying

jobs they can actually rely on. From Wyoming to Poland, and across Eastern Europe and Asia, coal communities are feeling hopeful and excited about this second lease on life.[15]

A Department of Energy study estimates a coal-to-nuclear shift could add over 650 jobs per site. And these jobs *pay*. According to the International Monetary Fund, nuclear investments have the biggest economic impact of *any* clean energy source. Nuclear industry wages average 36 percent higher than local wages, and nuclear workers make more per hour than their solar and wind counterparts. Plus, it's not just engineers and operators cashing in—these plants need everyone from custodians and security staff to accountants and electricians. If you're rooting for an energy transition that supports workers and their families, nuclear's got to be at the top of the list.[16]

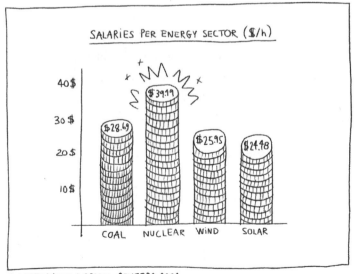

SOURCE: NICHOLA GROOM, REUTERS, 2021

IT'S NECESSARY

Every year countries meet at COP,* the United Nations' Conference of the Parties. It's essentially an annual festival where world leaders talk climate instead of dancing to DJs. In 2015, 196 parties got together at COP21 and signed the Paris Agreement, a historic deal to limit global warming to 1.5°C above preindustrial levels.[17] While it might seem like a random number, it's based on scientific evidence. The hotter the planet gets, the worse the impacts on ecosystems and human life. Limiting warming to 1.5°C compared to 2°C, for example, would mean fewer droughts and wildfires, which are becoming increasingly devastating to agriculture, forests, and communities. It's the difference between manageable changes and catastrophic impacts for many parts of the world.

The Intergovernmental Panel on Climate Change (IPCC), a group created by the UN, makes recommendations to cap that warming. The IPCC is there to guide everyone, helping countries set realistic goals, measure their progress, and tackle roadblocks. In 2022, it put out a report presenting four primary pathways to limit global warming to 1.5°C by the end of the century. Across *every* pathway, nuclear is seen as a key component to get there. Whether it plays a small or central role, the world's nuclear capacity must be at least *doubled* by 2050 to meet climate targets, according to the IPCC.[18] Sadly, 2024 was the hottest year on record and the first to top 1.5°C of warming. That doesn't mean we have failed to meet the Paris Agreement goal, because the tempera-

---

* COP is an international climate meeting held each year by the United Nations. The "parties" are the countries that have signed on to the UN Framework Convention on Climate Change (UNFCCC). Basically, these countries take turns hosting an annual meeting where government representatives discuss climate change and try to agree on ways to fight it.

ture increase needs to be consistent over decades instead of just one year. But it does mean we now have to fight even harder to stay on track. If we don't speed up the transition away from fossil fuels and drastically reduce emissions, we could be looking at 3°C of warming by the end of the twenty-first century, which would really, really suck.[19]

NUCLEAR IS NATURAL

Despite it being just another source of energy, people tend to think of nuclear as totally human-made and synthetic. It's *new* and *science-y* and way harder to trust than stuff that happens *naturally*, like fire. Right? Not so fast. It turns out the oldest nuclear reactor on the planet popped up nearly two billion years ago—and Earth made it all by herself.

In 1972, scientists discovered natural nuclear reactors.[20] You read that right. Rewind the clock 1.7 billion years and zoom in on the west coast of Africa, in a place we now call Oklo in Gabon.[21] The location featured caves with chunky deposits of uranium inside. Back in the day, when it rained, water filled these uranium caves, kick-starting the first ever reactors on Earth. Because of some special geological features we don't need to get into, the uranium atoms started splitting and sustained a chain reaction. If we could hop in a time machine and add a turbine and generator onto these all-natural nuclear reactors, they would have been able to make electricity.

Oklo produced an average of 100 kilowatts of heat energy on and off for about a million years—enough to power a thousand light bulbs.[22] That might not seem like much now, but remember, this happened without any human intervention.

Why does this matter? Apart from being one of the coolest fun facts ever, in a 2023 article for the Science History Institute, science writer Sam Kean made a fascinating point:[23] These reactors turned themselves on more than a billion years before fire was a thing on our planet.* Kean notes that we think of fire as natural and ancient—and relatively safe. Comforting, even. But nuclear is the OG source of energy on Earth. Just like fire, it can burn us if we're not careful. But it's every bit as natural as a flame.

Oklo also gives us valuable insight into what happens to nuclear waste over vast periods. The waste left behind by these natural processes, similar to what modern nuclear plants create, has only moved a few centimeters in the last billion-plus years. So, we know what could happen to nuclear waste very far into the future: not much.

NUCLEAR HAS THE LEAST IMPACT

Nuclear's footprint has all the other energy sources absolutely gagging. It's one of the cleanest, most efficient and environmentally friendly ways to make electricity. Don't just take my word for it; the United Nations Economic Commission for Europe (UNECE) dropped the receipts in a 2022 report. The report showed that nuclear has some of the lowest greenhouse gas emissions out of all energy sources. It also requires the least amount of land, mining, and metals. Hard to compete with that.

Nuclear really shows her stuff when you factor in something

---

* Fire needs oxygen, but terrestrial plants that fill the atmosphere with oxygen in sufficient amounts only started popping up on Earth around 440 million years ago.

called "life cycle analysis." I'll explain with the help of a topic that touches all of our lives: fashion.

Think about buying a pair of jeans. You probably know that some amount of water and energy went into creating them. Everything we do requires energy, after all. Levi Strauss, which is known for keeping tabs on its environmental footprint, says its jeans are responsible for around 73 pounds of carbon dioxide emissions a pair.[24] You can just imagine how much worse that number gets when it comes to fast fashion websites that don't show their numbers. Am I using this analogy to try to convince you to ditch fast fashion? Yes. It's like wearing coal.

Anyway, the footprint of your jeans isn't just about the water it took to grow the cotton or the electricity it took to weave and sew its fibers. It's not just about the gas used to ship it to your house either. When Levi Strauss does a life cycle analysis, it also looks at things like how often consumers wash their jeans—because that adds more water to the footprint—and how often they use hot water or a dryer, which require energy to produce heat.[25] They also care about washing habits because, as any denim aficionado will loudly and obnoxiously tell you, washing and drying your jeans is a great way to wear them out.

The point of this little fashion lesson is that the usage of energy and other resources isn't simple or static. There's the energy it takes to build the thing, the energy it takes to keep it running, the energy it takes to get it to the people who want to use it, the energy it takes to fuel it, and so on.

When it comes to this kind of cradle-to-grave environmental impact, nuclear electricity really shines. Its continuous and consistent output means fewer disruptions. This unwavering reliability not only ensures a stable electricity supply but also contributes to a more predictable and less volatile life cycle impact.

Nuclear is also compact, chic, and doesn't demand acres of land. Less land use equals less environmental impact—fewer forests cleared and fewer habitats disturbed.

Then there's nuclear's lack of decommissioning drama: Every diva has her last performance, and energy sources are no different. But not all finales are created equal. When it's time to retire, nuclear plants gracefully exit the stage. They're decommissioned and leave tiny environmental footprints behind. For all these reasons, nuclear has the lowest impact on ecosystems of all energy sources.

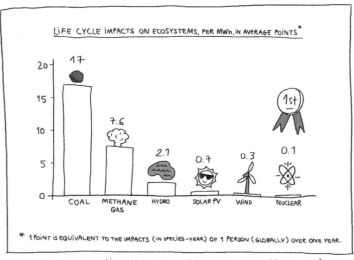

Reactors use a lot of water to cool down, so you might be worried that they'll drink us dry. Earlier in this chapter I talked about how nuclear power plants draw large volumes of water from rivers, lakes, or oceans, use it for cooling, and then return it right back to the source. Don't forget, this water doesn't touch the reactor core, so it's not radioactive. Some plants take it up a notch with cooling towers, reusing the same water over and over again. Still, it would be great to use even less water from natural sources. Thankfully, that's a possibility. Just look at Palo Verde in Arizona. It's the second-biggest nuclear generating station in the United States and nowhere near a natural water source. As a matter of fact, it's in the middle of the desert and pumps in treated wastewater for cooling. Side note: It's also possible to skip using water entirely and cool the reactors down using fans, though that's less efficient and a lot more expensive.

Nuclear reactors can in fact help us get *more* drinkable water. Their waste heat is ideal for powering desalination—turning salt water into stuff we can drink and water crops with. We already know it works; the Mangyshlak Nuclear Power Plant in Aktau, Kazakhstan desalinated 31 million gallons of water a day—roughly the water use of the entire city of San Francisco—when it ran from 1973 to 1999.[26] In 2021, researchers from MIT and Stanford showed that the Diablo Canyon nuclear power plant in California could be retrofitted to include a desalination plant. It would be able to make as much fresh water as the state's largest desalination plant in Carlsbad, but at half the cost per gallon.

TL;DR

There are magic spicy rocks in nature that can heat water to make steam. We've built machines that do this at scale; they're called "nuclear reactors." The really cool thing, besides using magic rocks to power our lights, is that this process doesn't release greenhouse gases or other nasty stuff into the air. It might sound human-made, but turns out Mother Nature was the first to go nuclear, with natural reactors popping up almost two billion years ago. I know. Wild. That's not just a fun fact to show off at your next dinner party. These natural reactors have the added benefit of telling us what happens to nuclear waste over very long time spans. As it happens, not that much. Investing in nuclear electricity makes our grid more reliable, creates loads of jobs, helps us protect the environment, and can even give old coal plants a new purpose—while giving us all the electricity we could ever want.

4: DISPELLING MYTHS

Despite the advantages we just talked about, you still might have the impression that nuclear is dangerous. You've heard about Chernobyl. You've heard about radioactive waste that lasts forever. As someone who has advocated for nuclear electricity for over four years, I know a thing or two about these fears.

In 2019, at the peak of my climate anxiety, I decided to do a ten-day fast in a clinic in Spain. Better health was the main motivation; little did I know that fasting is a spiritual practice that's been used historically for deep insights. After three days of pure misery, staring at my phone and saving recipes for when I could finally eat, clarity set in. I felt more energized and inspired than ever; my eyes were bright and shimmering with excitement. After the fast, I headed to a secluded cabin to do some more soul-searching and try to understand what role I could play in tackling climate change. One night, while brushing my teeth and staring at the mirror, a random thought popped in my head: What if I became a nuclear energy influencer? To be clear, I knew it was an insane idea. Nonetheless, I decided to run it by strangers next to me in lines or on airplanes, and their reactions showed me that this wild idea could have a big impact. They made me realize people didn't know anything about the topic and the little they did know was based on misinformation.

After months of research, I finally felt confident enough to share my message with the world. I created an alter ego called ISODOPE, an alien-looking avatar inspired by a medley of my favorite artists. ISODOPE made her debut on social media in March of 2020 holding a sign that read "Nuclear energy is clean energy!" To say that my followers, who were used to selfies and other model-type content, were shook would be an understatement. One of my modeling agents responded with "Nuclear energy isn't clean energy!" She later called, saying angrily, "It's a dangerous energy source and a contentious topic," also adding, "You are ruining your career." She hung up. She unfollowed. She never booked me for jobs again, and honestly, I was okay with that.

Even my friends started acting like I'd joined a cult. Some whispered about me behind my back. Others, bless them, were brave enough to say it to my face. "Why are you doing this?" "Your videos are weird." "Nuclear energy is bad." Cool. Thanks for the peer-reviewed feedback, guys.

By that point I had realized there are few issues more misunderstood and unfairly maligned than nuclear electricity. Let's clear the air on the misconceptions and set the record straight.

WHAT ABOUT CHERNOBYL?

In the twentieth century, humanity suffered its worst accident ever from an energy-generating facility. Hundreds of thousands of people died and millions lost their homes. If you think I'm talking about the Chernobyl nuclear power plant accident, you're wrong.

The deadliest energy accident in human history happened in China, and it was a hydropower dam collapse. The Banqiao Dam disaster eclipses every nuclear incident in history combined, but I'm willing to bet that you've never even heard of it. It certainly hasn't gotten the sleek HBO miniseries treatment, anyway. I'll have to set the scene for us.

It's August 1975, and Typhoon Nina is pummeling China's Henan Province with rain. Rivers are swelling and surging, setting the stage for an unprecedented catastrophe: the collapse of the Banqiao Dam, a feat of engineering dubbed "The Iron Dam." When it crumbled, the floodwaters swept away entire villages. The torrent of water from Banqiao and more rain caused another sixty-one dams across Henan to fail, leading to tens of thousands more people dying.[1] Some estimates place the total death toll as high as 220,000 due to famine and disease tied to contaminated water. Around 6.8 million homes were destroyed.

The 1986 Chernobyl disaster, in contrast, caused less than one hundred confirmed fatalities. No, that's not a typo. Less than one hundred. Of the 134 plant workers and emergency responders exposed to massive radiation doses, 28 died within the first three months after the disaster.[2] A further 19 died between 1987 and 2004 of various causes not necessarily associated with radiation exposure.[3]

This might come as a shock because you've been led to believe everyone involved met a gruesome fate. In the HBO TV series *Chernobyl*, one of the most dramatic scenes shows three plant workers going on a so-called suicide mission into the reactor's basement to drain water by opening a valve. The show implies they turned into radioactive toast. Not quite. While three workers did go into the basement, in reality they successfully completed their task and *survived*. Two of them were still alive and kicking in 2024, while the third passed away in 2005 from heart problems, almost two decades after the incident. So much for a "suicide mission."[4]

Tracking later deaths from radiation-related issues is much more complicated. In 2005, the World Health Organization estimated it to be around four thousand, but those numbers seem exaggerated. According to a report by the United Nations Scientific Committee on the Effects of Atomic Radiation (UNSCEAR), the only clear increase was in the number of cases of thyroid cancer. By 2018, over twenty thousand cases had been diagnosed among those exposed as children, with a survival rate of 99 percent. Apart from thyroid cancer, UNSCEAR found no evidence of increased rates of solid cancers or leukemia among the general population exposed to radiation from the Chernobyl accident.[5]

People often say that the confirmed death toll of the Fukushima disaster is around 2,300*—but that's kind of misleading. Those deaths weren't caused by radiation exposure due to the accident. The Japanese government has compensated one family for the

---

* When most people talk about the Fukushima disaster, they're referring to the meltdown at Fukushima Daiichi. Its sister plant, Fukushima Daini, was hit by the same earthquake but got by without any explosions or meltdowns. That's because Daiichi's failure was due almost entirely to mismanagement. You can assume I'm talking about Daiichi from here on out, unless I bring up Daini explicitly.

death of a plant worker from lung cancer, but there's no way to know for sure that the accident contributed to their sickness. All other deaths were due to things like overcrowded evacuation shelters and disruptions in hospital care. It's impossible to know which of those would have happened following the earthquake and tsunami that caused the accident, even if the meltdowns hadn't happened. In fact, we now know that an unnecessary fear of radiation contamination made the government evacuate more people than was needed.[6]

I'm not here to minimize the human suffering of Chernobyl or Fukushima Daiichi, and I'm definitely not here to demonize hydropower. Harnessing water for energy is great.

It's also remarkably safe. To put that into perspective, coal causes around 24 deaths per terawatt-hour* generated, mostly from air pollution.[7] Hydropower, by comparison, has a much lower rate of 1.3 deaths per terawatt-hour. Nuclear electricity is even safer, with only 0.03 deaths per terawatt-hour. Nuclear is just as safe as solar and wind.

One thing that Banqiao, Chernobyl, and Fukushima Daiichi all have in common? The energy source wasn't the problem. Yes, too much radiation can kill you; so can too much water. Both can *also* be harnessed safely and efficiently and are 99 percent safer than fossil fuels. There wasn't anything wrong with the technological concept of a massive hydropower dam *or* with the technology that powers nuclear reactors. The problem was that people—proud, greedy, careless people—cut corners.

---

* A terawatt-hour (TWh) is the cosmic unit quantifying an immense amount of energy—specifically, one trillion watt-hours. Imagine powering a city of millions for an entire year, and you're in the ballpark of a single terawatt-hour.

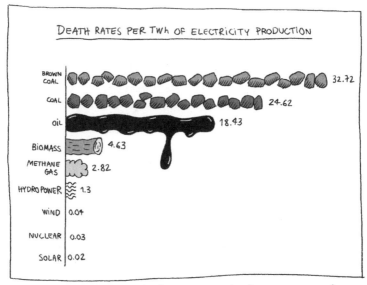

SOURCE: MARKANDYA & WILKINSON (2007); SOVACOOL ET AL. (2016); UNSCEAR (2008; 2018)

Now, let's not beat around the irradiated bush here: It's safe to say that Chernobyl was not a glowing success story for nuclear electricity. But avoiding nuclear because of Chernobyl makes as much sense as avoiding cruise ships because of the *Titanic*.

What actually went down? Could it happen again? Grab your hazmat suit and let's dive into the radioactive mess.

It all started with an experiment. In April of 1986, the operators of the Chernobyl Nuclear Power Plant were supposed to do a test in one of its four reactors to see how long the generator would continue making electricity after the reactor shut down.

The operators were about to reduce the reactor to low power— as it should have been for the test—when the central dispatcher of the Kyiv electrical grid called. It was just a few days before May Day, which was one of the USSR's biggest holidays. Factories were rushing to fill quotas before closing for celebrations, and a coal power plant nearby had gone offline. They needed all the electricity from Chernobyl. The operators decided to delay running the test until factory workers had gone home for the night and electricity demand had dropped.

Once people went to bed, operators rapidly took reactor four to low power. That's when things started going downhill.

Chernobyl's reactors had a design flaw: They had a positive void coefficient, which is a jargony way of saying they made more power when they got hotter. In most reactor designs, power production goes down as things heat up.

We'll skip through about a thousand pages of complicated nuclear physics to get to what happened. The reactor got hotter and hotter,* leading to an uncontrolled chain reaction and an explosion. The blast blew the lid of the reactor through the roof, releasing a lot of radiation.

The carelessness continued as the USSR responded to the accident. Just to hammer home how shady this was, the rest of the world didn't learn about the disaster until *days* after it happened. Two days after the explosion, radiation alarms went off at a nuclear power plant in *Sweden*, over one thousand kilometers away from Chernobyl. After thorough checks, Swedish workers found no issues with their reactors.

Puzzled by the alarm, they examined the radiation patterns and figured out that the contamination was coming from somewhere else. By analyzing the wind and the type of radioactive isotopes, they traced the origin back to the Soviet Union. This forced the Soviets to disclose the disaster to the world, several days after the explosion had already caused radioactive contamination.

Even though Chernobyl was considered the worst nuclear elec-

---

* Side note: No one makes reactors this way anymore. Lesson learned.

tricity disaster in history, it had a much lower impact on human health and the environment than was anticipated. The area didn't even turn into a barren wasteland as it's generally portrayed. On the contrary, the exclusion zone is now the third-largest nature reserve in mainland Europe and has turned out to be an accidentally successful experiment in rewilding.[8] With fewer humans around, the area has turned into a wildlife refuge, with lynx, bison, deer, and other animals roaming freely through the forests. There are also an estimated two hundred people who ignored safety warnings and went back to live in their villages located in the exclusion zone. Most of them are now in their nineties.

Here's the most mind-blowing fact: One of Chernobyl's other reactors (obviously not the one that exploded) kept operating until the year 2000. This isn't to say "Yay, Chernobyl!" but just to highlight how, when it comes to nuclear, the reality is always different from public perception.

Then there's Fukushima. On March 11, 2011, a 9.0 magnitude earthquake struck off the coast of Japan and created a powerful tsunami. The waves peaked as high as 131 feet and inundated the Fukushima Daiichi Nuclear Power Plant.

The tsunami knocked out the electricity supply and backup generators, causing a complete loss of power to the plant. Without power, the cooling systems stopped working. Three of the plant's reactors experienced a meltdown.*

The Japanese government declared a twelve-mile exclusion zone and rapidly evacuated tens of thousands of residents. This

---

* Just how melty is a meltdown? Here's the 411: Meltdowns happen when a reactor's cooling systems fail. As the temperature increases, the fuel rods containing the nuclear fuel may begin to melt, releasing radioactive materials. Despite what some TV shows and movies might tell you, there's no scenario in which the reactor melts down into the core of the planet.

did not go well. Overcrowded evacuation centers and disruptions in medical care were entirely responsible for the death toll of 2,313 people.

The unprecedented scale of this disaster made people feel like nuclear electricity must be risky. But like hydropower, it's not inherently dangerous. After the Banqiao Dam failure killed tens of thousands, no one called for abandoning hydropower. Instead, efforts focused on improving safety. So why treat nuclear differently?

Even the tsunami that led to the Fukushima Daiichi meltdown didn't *have* to turn into a nuclear disaster. After a safety test in 2008, the folks in charge were told that the plant needed a higher seawall. The test showed that in the event of an earthquake, the plant could get hit with a wave as tall as fifty-two feet. The plant's seawall was only nineteen feet high. That's . . . a lot less than fifty-two. The obvious recommendation was to add more protection, but an executive nixed the idea. Tragically, regulators brought this back up just *four days before* the earthquake that led to the accident. By then, of course, it was too late. If the plant had built up its defenses in 2008, the 2011 tsunami might not have flooded the diesel generators. If the backup power had stayed on, the cooling system would've kept running. No meltdown, no problem.

A lot was learned from Chernobyl and Fukushima. Every time there's been an accident at a nuclear power plant, the entire industry has leveled up its safety game. Because of Fukushima, plants in the United States now have a robust system to prevent a similar situation. We're talking extra diesel generators, hookups for fire pumps, and a whole suite of other tools for dealing with natural disasters and other accidents that could interfere with normal operations.

Perhaps more importantly, these incidents taught us that how we prepare for and react to accidents makes all the difference. At Banqiao, people ignored reports and kept anti-flood infrastructure to a minimum. At Chernobyl, engineers who ran the reactors were kept in the dark about the facility's design flaws. The authorities then wasted precious time denying how bad the disaster was. At Fukushima, they ignored data on the risk of tsunamis in the area and failed to implement proposed safety measures or make plans for the mass evacuation that, when fumbled, killed so many.

Reading about these accidents makes nuclear sound scary. But in more than seventy years, there have only been a handful of commercial nuclear power plant accidents that resulted in a release of radiation. Only two of them caused any fatalities. Even when you factor in these rare incidents, nuclear electricity's risks still pale in comparison to the daily damage done by the normal operation of fossil fuels. In the best-case scenario, without any accidents or spills, pollution from burning fossil fuels alone causes heart and lung conditions that lead to millions of deaths every year.

If you cared about protecting human life, you would be nuclear's biggest cheerleader. It's not dramatic to say that this exaggerated fear is killing us. After Germany shut down its nuclear plants in response to the Fukushima disaster, coal was used to replace the lost power. A study found that this shift has resulted in at least 1,100 premature deaths every year due to the increase in air pollution.

Another study estimated that nuclear electricity saved around two million lives between 1971 and 2009 by cutting down on air pollution.[9] But I guess it's hard to make a hit TV show about that.

WHAT ABOUT RADIATION?

At risk of stating the obvious: Radiation is scary. It's something we can't see, smell, or sense that has the *potential* to kill us. You might have heard that there's "no safe dose," but it's not that simple. We are exposed to low doses of radiation *all the time*. The human body itself has radioactive elements and minerals in it. Each year, the average American gets a dose of around 6.2 millisieverts of radiation.* That's about as much as you'd get from eating 62,000 bananas.† [10]

Half of that comes from natural, unavoidable sources, like radon gas in the air and cosmic radiation from the sun and beyond. The other half comes mostly from medical procedures like X-rays, plus foods with naturally occurring radiation like bananas and Brazil nuts. None of this is harmful. Bananas and Brazil nuts are great.

Some people are exposed to higher-than-average levels of radiation, without any problems. Folks in Denver, for example, get about 1 extra millisievert a year just from living a mile above sea level.[11] That's because there's less atmosphere up there to shield them from cosmic rays.

Being in a plane increases exposure as well. In fact, commercial airline workers receive about twice as much radiation as nuclear plant employees per year.

In contrast, the average person gets about one thousandth of a millisievert from nuclear power plants. Even if you live right next door to one, the amount of radiation you're exposed to in a year is the equivalent of about eighty-one seconds on an airplane. During

---

* I really, really wish there was a better way to explain what a millisievert (mSv) is. The answer is that it's a thousandth of a sievert (Sv). This unit is used to measure not just how much radiation energy hit your body, but how dangerous that energy is. It's based on how much energy (in joules) gets absorbed per kilogram of tissue. But it also multiplies that by a factor depending on the type of radiation, because some kinds are way worse for you than others. I'm sorry!!!

† Yup, bananas are radioactive. That's because some of the potassium that makes them so good for you comes in a radioactive form. So many people use bananas to put low doses of radiation into context that there's an unofficial unit called a "banana equivalent dose," or BED.

the biggest nuclear accident in U.S. history—Three Mile Island,* which released some radioactive gases in 1979—the max dose to a member of the public was 0.8 millisievert. That's less than one sixth of the yearly dose on average, or about 8 percent of a stomach CT scan.[12]

---

* We will talk about Three Mile Island in chapter 7.

What about accidents? It's true that getting hit with a big, sudden dose of radiation is very, very bad for you. Then again, so is being electrocuted. Or getting hit with a massive wave of water.

In 1954, the National Council on Radiation Protection and Measurements tried to set maximum safe doses of radiation. Here's the problem: As radiation levels drop to the kinds of exposure you get from nuclear reactors, the risk gets so teeny-tiny that it becomes impossible to study. You can't isolate all the variables that introduce us to even *more* radiation on the regular, like X-rays and air travel. That's why scientists created something called the "linear no-threshold model" (LNT). It uses statistics to try to figure out the risks of low doses of radiation. It takes what we *do* know—that very high doses cause death or increase cancer risk—and uses that to guesstimate what lower doses might do. It's a "better safe than sorry" mindset. The model predicts that any amount of radiation has the potential to cause an increased risk.

Naturally, we've become terrified of radiation. There's even a word for this specific brand of terror: "radiophobia." This fear doesn't just make people less likely to support nuclear; it can be a lot deadlier than radiation itself. It can lead to the refusal of lifesaving cancer treatments and important diagnostic scans that use radiation. Strangely enough, countries with high levels of radiophobia, like Germany and Austria, also have radon gas spas where people willingly expose themselves to low levels of radiation for supposed health benefits. In some instances, these treatments can even be covered by health insurance. Make it make sense.

The real world tells us a different story about the dangers of radiation. Ramsar in Iran has some areas with radiation levels as high as Mars's surface (super radioactive). People in this town can receive an annual radiation dose of up to 25 rem (25,000 mrem), which is five times higher than the 5 rem allowed for American radiation workers each year. It's only logical to conclude that the residents of Ramsar have a higher incidence of cancer or birth defects, right? Nope. Studies show no apparent increase in birth defects, no increase in rates of cancer, and no reduction in lifespan compared to Iranians in other areas.

It's also important to note too that people who freak out about radiation *don't really care about it*. Because if they did, they'd embrace nuclear with open arms. Turns out that a coal power plant producing the same amount of electricity as a nuclear power plant will release at least *ten times more radiation* into the environment (though radiation is not what makes people sick, it's all the other stuff).*[13]

The outsize fear of radiation was written all over the public fallout from the Chernobyl accident. Many historians and public health experts now argue that while of course some people did suffer terrible health impacts, the effects of fear and misinformation were even more damaging. And we're still spreading a lot of those myths. Just look at how wildly successful the 2019 HBO miniseries *Chernobyl* was. HBO clearly did a ton of work to nail the aesthetic vibes of the era, but they absolutely dropped the ball on the science.

The most emotionally compelling storyline features the pregnant wife of a local firefighter. While it's true that the first responders were the ones who got horribly sick—unlike pretty much everyone else, despite what the miniseries shows—things completely go off the rails from there. Doctors keep warning the woman that she can't be around her husband because he's super-*duper* radioactive.[14] In fact, the main reason people with acute radiation

---

* Coal has small amounts of natural radioactive elements that can become more concentrated when it's burned, adding to environmental radiation.

syndrome are isolated is for *their own* protection, because their bodies are shutting down and they're incredibly susceptible to infection. You can't catch radiation poisoning from being in the same room as someone who's contaminated, after they've removed their clothes and showered. Even in the case of these workers—who were in a unique situation due to directly handling extremely radioactive materials—the danger would be in hugging them or handling their dirty clothing, not in keeping them company.

We're later told that this character—who demanded to stay by her husband's side—gave birth to a horrifically ill baby. "The radiation would have killed the mother, but the baby absorbed it instead," a scientist says, very unscientifically. "We live in a country where children have to die to save their mothers."

I actually want to thank HBO for giving us an example of the dangers of radiophobia that's almost *too* perfect. Because there's absolutely zero evidence of Chernobyl causing birth defects. Seriously. *Zero evidence*. Even if you hugged someone covered in alpha emitters and got contaminated, there's no way for a fetus inside your body to soak up all that radiation like a sponge.

After the disaster, fears of radiation were so widespread that women even terminated healthy pregnancies. Robert Gale, a UCLA physician who coordinated the international medical response to the accident, told newspapers in 1986 that Soviet doctors were advising pregnant women to have abortions.[15] While officials claimed the decisions were voluntary, some survivors later described feeling pressured. In 1987, researchers estimated that up to two hundred thousand *wanted* pregnancies were ended in Western Europe because doctors told patients that radiation from Chernobyl posed a risk to the fetus.[16] Abortions spiked as far away as Denmark and even Greece.[17]

Tragically, Gale himself concluded in 1987 that Chernobyl

hadn't had any impact on growing fetuses. "We've now had a chance to observe all the children that have been born close to Chernobyl," he told The New York Times. "Not surprisingly, none of them, at birth at least, have any detectable abnormalities. We weren't expecting any."[18]

Scientists also kept a close eye out for the much-hyped uptick in cancers people anticipated. In practice, only cases of thyroid cancer went up. Because the government was slow to tell people about the accident, they unknowingly ate and drank contaminated food and milk.[19] Fortunately, that form of cancer responds very well to treatment. The survival rate of folks with Chernobyl-adjacent thyroid cancer is estimated to be 99 percent.

Sadly, life expectancy among survivors of the incident did drop, though this was mainly due to alcoholism and suicide.[20] We're seeing more of the same in the aftermath of Fukushima Daiichi, where loads of people have lost faith in their government because of how the accident was handled.[21]

WHAT ABOUT URANIUM MINING?

It's impossible to talk about nuclear electricity without talking about uranium mining—as it should be. We literally gotta mine uranium if we want to have nuclear, and it's a complex process with a sketchy history. Yet, like most things relating to nuclear, the talking points people bring up about uranium mining are super tired.

The truth is that uranium mining has become one of the safest, most tightly regulated kinds of mining in the world.[22] This is for the very same reason that many environmentalists and human rights advocates turn their noses up at it: because it used to totally, completely suck. Let's take a little trip back in time to see where it all went wrong, how it got better, and why the general public's perception of uranium mining is stuck in the 1950s.

Uranium was discovered long before nuclear fission. It first turned up in Czech silver mines in 1789, tucked inside the bubbly-looking mineral called "uraninite," also known as "pitchblende." Miners quickly realized it was something special, in part because they seemed to get kinda sick after being around it. (Side note: Most of the risk from uranium mining comes from inhaling it in dust form, not from touching it.) The mineral quickly became famous for creating bright yellow colors and green fluorescence in glass. In 1895, we figured out it was radioactive, and the rest is history.[23]

For the first century or so after uranium's discovery, it was mined in a free-for-all, like the gold rush—dudes in dirty blue jeans scratching their butts and panning rivers—except they were handling radioactive rock dust instead. This was, to state the obvious, not a great way to mine uranium. The General Mining Act of 1872 meant that companies could grab up federal land for a small fee if they discovered "a valuable deposit" of minerals there. Many mining towns were abandoned without any attempt at cleanup once demand dropped. Today, there are at least fifteen thousand abandoned uranium mines on U.S. federal lands. That might sound like a lot, but there are an estimated five hundred thousand abandoned mines of all sorts in the country. Many of them are just as nasty, if not worse.

As demand for uranium picked up during and after World War II, military-fueled mining initiatives prioritized speed and efficiency above all else. So, even though the government was now paying attention, they weren't doing much of anything to protect workers or the environment. Most mines didn't have good ventilation or dust control. As miners huffed and puffed their way through their shifts, they were basically hotboxing radioactive gases. To make things worse, mines in the fifties regularly left their waste piled up outside, where it threatened people and animals. Uranium dust could get stirred up by the wind and travel into nearby homes and water sources. As time goes by, uranium decays into radon gas, which makes it more dangerous. By the way, uranium from the earth naturally turns into radon gas all the time. That's why it can build up in your basement and make you sick.

Uranium mining also often caused disproportionate harm to Indigenous communities. One particularly horrifying example is the Navajo Nation. The Navajo people opened their land to the U.S. government during World War II. The government was secretly after the area's uranium deposits to support nuclear weapon development. Between 1944 and 1986, miners extracted 30 million tons of uranium from the Navajo Nation and abandoned at least five hundred mines.[24] Because they weren't told about the risks, residents swam in contaminated water and even used uranium rock to build their homes.[25]

One study being done in collaboration with the Centers for Disease Control and Prevention has found that 27 percent of Navajo Nation participants still have high levels of uranium in their urine, compared to 5 percent across the total U.S. population.[26] Despite having relatively low risks of many cancers and low smoking rates, people in the Navajo Nation have unusually high rates of lung cancer.[27]

This legacy is devastating, and the communities affected by past uranium mining practices deserve full acknowledgment and proper compensation.* Thankfully, modern uranium mining is much safer and less messy. But let's not pretend mining is ever totally

---

* They are starting to be compensated. The Environmental Protection Agency (EPA) has received funding to assess and remediate 230 of the 523 mines and has invested as much as $5 million per year toward investigation and cleanup of contamination and technical support for Navajo government agencies.

harmless. Mining *anything*—gold, coal, rare earth metals—has a habit of wrecking the environment and messing with people's lives, especially Indigenous communities.

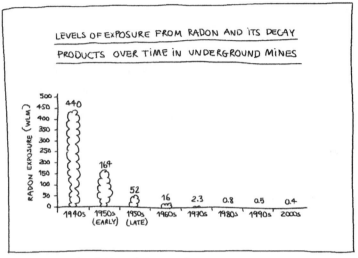

SOURCE: AREVA (2011), KIGGAVIK PROJECT ENVIRONMENTAL IMPACT STATEMENT, AREVA RESOURCES CANADA INC., SASKATOON

MINING SUCKS

A lot of people hear "mining" and picture scruffy men with pickaxes covered in coal. But mining isn't unique to fossil fuels and nuclear, and unfortunately mining sucks. Gold mining in Peru is considered responsible for the deaths of almost half of the Indigenous Amarakaeri people. Lithium mining in Chile is currently contributing to severe drought and displacing people from their ancestral homes.[28]

Switching to renewables and batteries won't make mining go away either. A single electric car battery requires around 35 pounds of lithium and 100 pounds each of nickel and manganese.[29] All that stuff has to be mined. Cobalt is another crucial component in many electronics' batteries. Most of the world's cobalt is sourced from the Democratic Republic of the Congo, where ethical concerns loom large. Reports from Amnesty International and other watchdog organizations show that rising cobalt demand has worsened the country's rampant child labor, dangerous working conditions, and environmental degradation.[30] Studies suggest that working in cobalt mines raises the risk of numerous health problems and birth defects.[31] Demand for cobalt is expected to reach more than 222,000 tons by 2025, which is three times the demand seen in 2010.[32]

There are dozens of minerals and metals we're going to need more of as we transition away from dirty energy sources—there's no sugarcoating that. It doesn't mean we should delay switching to solar panels, batteries, and nuclear reactors though, because they're still a huge improvement over fossil fuels. Hannah Ritchie, a data scientist and senior researcher at the University of Oxford, compared the numbers and found that even if we go all in on clean energy as fast as possible, the amount of mining needed will still be five hundred to one thousand times less than the mountain of stuff we currently dig up for fossil fuels.[33]

By 2040, we'll need to mine 28 million metric tons of materials for clean energy. While that might sound significant, it's nothing compared to the 15 *billion* metric tons of coal, oil, and gas we currently burn through every year.[34]

Nuclear's incredible energy density means way less mining compared to other clean energy sources. The Breakthrough Institute found that producing one gigawatt-hour of electricity with nuclear requires mining only 30 percent of the rocks and metals needed for solar and just 23 percent for onshore wind.[35]

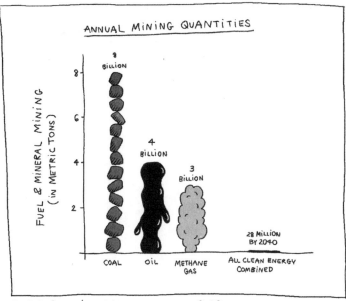

SOURCE: SUSTAINABILITY BY NUMBERS, 2023

Even though clean energy tech requires far less mining than fossil fuels, this doesn't mean we should sit back and call it a day. There's still plenty of room to improve supply chains across the board, and those industries could learn a thing or two from uranium mining, which has largely gotten its act together.

These days, more than 50 percent of uranium is mined using in situ recovery.[36] This process skips the need for open pits or tunnels. Instead of blasting or digging, a liquid, typically made of water, oxygen, hydrogen peroxide, and either sodium carbonate or carbon dioxide, is injected into the ground to dissolve uranium, which is then pumped to the surface. No heavy lifting, no giant holes in the ground, and way less mess. Even old-school tunnel mining, still used in some areas, now relies mostly on machines, sparing workers the joy of shoveling rocks all day.

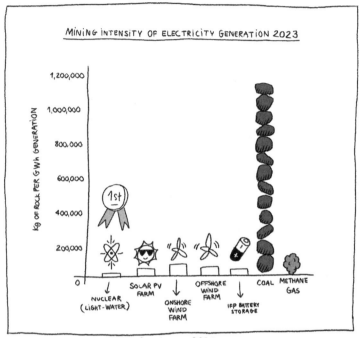

SOURCE: THE BREAKTHROUGH INSTITUTE, 2024

So yes, it's true that uranium mining has been a total nightmare in the past. But this isn't just a uranium problem; it's an everything-we-dig-up problem. The injustices faced by the Navajo Nation should remind us to consider ethics whenever we mine *anything*. As we move forward, the focus should be on equitable partnerships with local communities, robust environmental rehabilitation, and a commitment to transparent, responsible practices.

Do I wish we could wave a magic wand and never mine anything again? Absolutely. Would it be great if our clean energy tech had squeaky-clean supply chains and glowing résumés? Obviously. But here's the harsh reality: Perfection doesn't exist, and holding out for it is just a self-righteous way of doing nothing. There's a whole army of armchair critics out there, lined up and ready to scream "Not good enough" at every possible solution. While they nitpick every innovation to death, the world keeps heating up, air pollution keeps killing millions, and countless species inch closer to extinction.

The best we can do is keep improving our technology and reducing its impact step-by-step. Progress isn't perfect; it's messy, incremental, and full of compromises. But you know what's worse than messy progress? No progress.

WHAT ABOUT TRITIUM?

Now, let's spill the tea on water contamination fears. Some people worry that nuclear plants releasing cooling water into rivers or oceans is a disaster waiting to happen. They worry about potential radiation leaks and generally get the ick about having nuclear stuff touch their water supply.

When people freak out about nuclear water contamination headlines, the culprit is usually a radioactive isotope* of hydrogen called "tritium." Tritium is all around us. It's used as the secret sauce for things like glow-in-the-dark exit signs and watch hands. It also occurs naturally; we're exposed to it on a daily basis.[37]

While the water that cools reactors doesn't come into contact with the fuel itself, some of the neutrons that make fission happen *can* escape the core and bounce into the cooling water. That's normal, expected, and planned for. When these neutrons hit the

---

* Just a refresher: An isotope is a variant of a chemical element. Just like how ice cream can have different flavors but still be ice cream, an element can have different isotopes but still be that element. Hydrogen has three isotopes, tritium being the only radioactive one.

water, they sometimes collide with the hydrogen in it. If a neutron manages to stick to a hydrogen atom, the atom transforms into a slightly heavier version called "deuterium" (basically, hydrogen with one neutron attached). If another neutron comes along and crashes into that deuterium, it turns into tritium—the radioactive cousin of hydrogen.

This process doesn't just happen in nuclear reactors; it also happens all day long in the atmosphere with neutrons that come from space.[38]

Some folks get antsy about tritium going into water sources. I get it, we're evolutionarily wired to freak out about contamination. But in the world of radioactive isotopes, tritium is about as harmless as you could ask an isotope to be.

Tritium emits a type of radiation that's very low energy compared to other types of radiation. This means the radiation it gives off can only travel a short distance and cannot penetrate human skin. If tritium is ingested (usually in the form of tritiated water), the human body processes and eliminates it very quickly. Its biological half-life is about ten to twelve days, meaning half of it is flushed out in that time through your pee or sweat.

The EPA and the World Health Organization (WHO)—who, let's be honest, have seen some things—have both looked into tritium and concluded it's not a health risk at the levels typically found in the environment.[39] You'd need to guzzle down a swimming pool's worth of the most tritium-contaminated water in one go to get anything close to a dangerous dose. And I don't know about you, but I already struggle to drink eight glasses a day.

Nuclear power plants keep close tabs on the water they pump in and out of their cooling systems. They know exactly how much tritium they're releasing, and scientists have confirmed time and time again that those levels are safe.

Tritium made splashy headlines after the Fukushima accident. Since 2011, the Japanese power plant company TEPCO has been pumping water to cool down the damaged reactors. Unlike the water used to cool a healthy and thriving nuclear reactor, this water does directly come into contact with the radioactive core. For over twelve years, TEPCO has been storing the contaminated water in tanks and there's now more than enough to fill five hundred Olympic swimming pools. But that water can't stay there forever.[40]

TEPCO is planning to release it into the ocean. And no, it's not just being dumped. The water gets run through a high-tech filtration system that removes sixty-two different radioactive isotopes.[41] However, these filters can't remove tritium because it's part of the water itself. As a special type of hydrogen, tritium bonds with water molecules to form tritiated water, which behaves just like regular water. So instead, they do the next best thing: they dilute the water until the tritium levels are so low they're basically negligible. The water being released from Fukushima has six times less tritium than the World Health Organization allows in safe drinking water.[42] In other words, you could technically bottle it, slap a trendy wellness label on it, and sell it to LA influencers. (But maybe don't.)

In 2023, with the approval of the International Atomic Energy Agency (IAEA), TEPCO started to very, very slowly release this filtered and diluted water into the nearby ocean. Like, seriously slowly: It's going to take thirty years to clear it all out.

I'm not too concerned, because there's significant outside testing going on to make sure the water, sediment, and fish aren't impacted. The IAEA, an international organization dedicated to keeping nuclear safe, oversees the whole process. They even have a page on their website where you can see the live data, including the concentration of tritium and the rate of release. No need to just take TEPCO's word for it—you can check it all yourself.

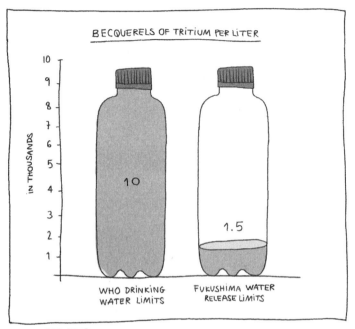

SOURCE: EPA, 2023

But what about eating the fish from the waters around Fukushima? Researchers from Oregon State University crunched some numbers, and you'd have to eat nothing but the most contaminated tuna for 578 years to ingest the same amount of radiation you're naturally exposed to in twelve months.[43] If you're counting on living that long *or* eating that much fish, I think you've got bigger fish to fry.

WHAT ABOUT NUCLEAR WASTE?

I want you to imagine nuclear waste in your mind for a sec. Let me guess . . . rusty barrels stacked in a creepy warehouse, glowing green goo dripping dramatically from cracks, maybe even some fog rolling in for effect. Classic. But that's Hollywood fiction.

In the real world, nuclear waste is incredibly boring and uneventful. Imagine rows of massive concrete storage containers lined up in a quiet fenced-off area under constant surveillance. No glowing, no goo, and definitely no toxic sludge oozing its way toward a town. Just the safest, most tightly managed waste you'll ever come across. This is what it looks like:

You've probably heard the horror stories: Nuclear waste is unbelievably dangerous, stays radioactive for hundreds of thousands of years, and we have absolutely no idea what to do with it. But sometimes things sound more dangerous than they actually are.

When people bring up these concerns, they're talking about "spent nuclear fuel," which is the technical term for "nuclear waste." In reality, it's not even waste. It's just fuel that's already done its job in a reactor and is hanging out, waiting for its next chapter. In my opinion, spent fuel isn't a problem we're ignoring; it's a gold standard for waste management. Every ounce is logged, stored in ultra-secure containers, and babysat 24-7 like it's the crown jewels. It doesn't leak, wander off, or decide to randomly blow up. In fact, spent nuclear fuel might just be the most well-behaved "waste" product on the planet.

Let's get down to the details. What is it? How is it handled? And why is it a shining example of waste management?

Every 1.5 to 2 years, nuclear reactors have to get fresh fuel in and take used fuel out.* From the outside, this used fuel looks exactly the same as fresh fuel. It's a bunch of solid ceramic pellets, lined up neatly inside long metal rods like a radioactive Pez dispenser. The difference is that spent fuel is extremely radioactive and hot and needs to be handled properly.

The first step is to take it out of the reactor's core and put it in a deep pool, called (very uncreatively) a "spent fuel pool." Engineers are great at safety, not branding. The water does two things: It cools the spent fuel down and blocks radiation. As *XKCD* comic writer Randall Munroe explains in his book *What If?* this method is so effective that you could swim in a spent fuel pool without getting hurt.[44] Not that you should. You'd almost certainly get arrested, possibly tackled by very serious men in hazmat suits. But scientifically speaking, the water shields you so well that you'd have to do something incredibly dumb, like free dive to the bottom and cuddle up to the fuel rods, to get a harmful dose. The point isn't that you should practice your diving skills in a spent fuel

---

* Over time, fission products build up in the fuel, absorbing neutrons and slowing the chain reaction that powers the reactor. While most of the uranium remains intact, the fuel becomes less efficient and has to be replaced regularly to keep the reactor running smoothly.

pool; the point is that we know how to protect humans from nuclear waste quite easily, even during its most dangerous stage.

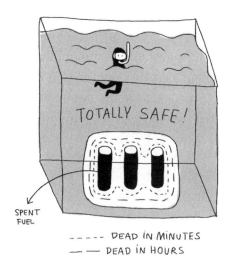

After about five years, the spent fuel is transferred into massive concrete and steel containers called "dry casks," which look like giant cans of Pringles. They're designed to be able to handle virtually anything. I mean *anything*. People have crashed trains into them, set them on fire with jet fuel, and more—you can even find videos of these tests online. Once sealed, the casks are placed on a concrete pad at the nuclear power plant alongside others, where they simply sit, quietly doing nothing. And that's essentially the end of the story. In case you're wondering, no one has been injured or killed by commercial nuclear waste stored in dry casks. Not a single person.

This is where nuclear naysayers face an inconvenient truth. This is the only industry, perhaps on the planet, that knows where every single gram of its waste is. Dry cask storage has proven to be a safe and easy way to manage nuclear waste, at least in the short term. It's so effective that you can hug one of them without

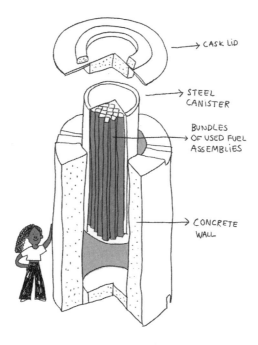

any concern about excess radiation exposure. Again, not recommending this, as you will most likely be arrested.

But going from the reactor straight into a pool and later into a dry cask is not the only option for disposing of nuclear waste. Some countries, like France, can recycle it. Around 17 percent of France's electricity comes from reprocessed spent fuel.[45] Not all countries bother reprocessing spent fuel because, honestly, it's just cheaper to buy fresh fuel.

Nuclear waste becomes safer over time, unlike things like mercury, which stays dangerous forever. Only a tiny fraction of the waste remains hazardous for thousands of years—and even then, it's only a problem if people somehow ingest it. Turns out, we have been great at isolating spent nuclear fuel and not letting it contaminate anything.

That said, dry casks are only licensed for about a hundred years, and spent fuel needs to stay safely separated from people and the environment for much longer than that. One approach is to stick with dry casks indefinitely, replacing the old ones with new ones every century. It's straightforward, manageable, and doesn't require any drastic changes.

I understand it can feel uncomfortable to think about very long-term waste management. But let me put it into perspective: The United States creates about as much mercury waste as spent nuclear fuel in a year.[46] Most of this waste comes from burning things like coal and wood and goes into the atmosphere. Mercury is extremely toxic, is brain-damaging, and doesn't even glow in the dark. And it needs to be kept away from people and the environment *forever*. Yet, you don't hear people losing sleep over mercury waste management thousands of years down the road.

Now consider plastic. Plastic is the final boss of the modern world and it *will* quietly haunt you for centuries. We're currently cranking out 400 million tons of plastic per year, an absurd amount, and most of it ends up in oceans, landfills, your sushi, your bloodstream, and now apparently your balls. In 2024, scientists found microplastics in 100 percent of human testicles they tested. If you're lying awake at night worrying about a few thousand tons of nuclear waste that's safely locked up and hasn't hurt a soul, maybe pivot that anxiety toward the mountains of trash, clouds of mercury, and microscopic plastic confetti currently turning our organs into a science experiment.

But maybe you're afraid we don't know what to do with nuclear waste in the *super* long term. Like, thousands of years down the road. Which is ironic, considering if we don't get off fossil fuels, we might not make it that far.

Plenty of folks worry that dry casks won't be enough to protect humans from nuclear waste if, say, society as we know it collapses. There's a whole field dedicated to figuring out how we could communicate the waste's danger without using English, in case people in the future can't understand it. It's called "nuclear semiotics," and its scholars have suggested everything, including putting literal cats that glow in the dark in the waste facilities. Because future humans will be able to decode danger through interpretive feline symbolism or . . . something? Another idea is to put up signs that say deranged stuff like "This place is not a place of honor . . . no highly esteemed deed is commemorated here . . . nothing valued is here," which sounds more like a journal entry from someone deep into an existential crisis, and has inspired this great meme:

Dispelling Myths

While we haven't decided how best to warn postapocalyptic humans or confused alien visitors, scientists agree that burying nuclear waste deep in stable rocks is a solid and safe plan. Remember, they're not just pulling this idea out of thin air. We know what happens to nuclear waste over long stretches of time because of Earth's natural nuclear reactors. Unlike modern waste facilities, Oklo didn't have any engineered barriers to prevent that waste from spreading far and wide—it just stayed put.

Finland is leading the charge with the Onkalo deep geological repository, the world's first to come online. And if digging a fancy underground cave isn't extreme enough for you, there's also the deep borehole option, which involves drilling a hole about three

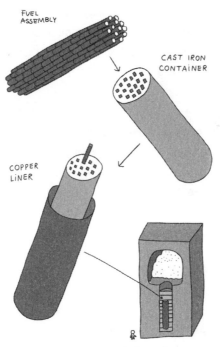

Onkalo deep geological repository

miles deep and stacking spent fuel inside before closing it off, isolating the waste even farther from the environment. Then again, we might not want to put this waste away forever. Remember reprocessing? Only a tiny percentage of spent fuel is stuff we *definitely* don't want. Over 94 percent of it could still be used to make electricity. That's like tossing your entire Thanksgiving leftovers because you don't want to eat the cranberry sauce.

All this waste will be tucked away in rock formations that haven't moved in millions of years. It will be far away from any water tables and locked up tightly in highly specialized casks.

I'd simply love to know how a civilization that's forgotten how to read or decode human language is somehow going to reinvent industrial-grade excavation equipment *just* to crack one of these things open. That's a level of curiosity and determination I frankly admire. And if accidentally exposing postapocalyptic alien visitors to nuclear waste is your main argument against using nuclear electricity . . . You know what? I'm not even going to finish that sentence.

Now, here's the real kicker: If all the electricity in the United States came from nuclear, each person would create just 34 grams of spent nuclear fuel per year.[47] That's less than the amount of sugar in a single 12-ounce Coca-Cola. If you powered your *whole* life with nuclear—every road trip, scroll session, and laundry load—all the spent fuel you *ever* created would fit inside a soda can.

Compare that to more than 10,000 kilograms of CO_2 per year from coal for the same amount of electricity. In fact, coal plants churn out thirty-two times more waste every single day than every U.S. nuclear plant has produced in forty-five years.

We've already talked about the countless lives lost to fossil fuels' waste due to air pollution: more than four million of them *each year.* But even environmental darlings like solar produce some

amount of hazardous waste. As we embrace the sun's power, we're also creating a mountain of discarded panels. The International Renewable Energy Agency estimates that by 2050, we'll have about 78 million metric tons of old solar panels to deal with.[48] They're tricky to recycle with current tech and often contain dangerous metals like lead that can seep into the soil. In 2024, there are no federal regulations in America for dealing with old solar panels or wind turbines, which tend to end up in the landfill.

Of course, all of this pales in comparison to fossil fuels. And, like spent fuel from nuclear reactors, solar panels and wind turbines *could* be recycled (and already are in some countries).

None of this is to say we should be *casual* about nuclear waste. But it's also not the terrifying sludge it's so often made out to be. The point isn't that we should be worrying *less* about nuclear waste. It's that maybe we should be worrying *just* as much about every other kind of waste we seem weirdly chill about.

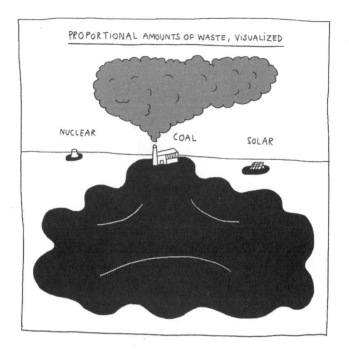

TL;DR

We stopped using the magic rocks that heat water because they exploded that one time. Imagine if prehistoric people stopped using fire because a couple of guys burned their houses down. Nuclear reactors are safer than driving. Chernobyl wasn't even the worst energy disaster in history (as a matter of fact, one of the plant's reactors was still kicking and making electricity until the year 2000). Nuclear waste in dry casks is a nonissue. Uranium mining used to suck, but it's tightly regulated now. Turns out all the nuclear fearmongering you've heard is mostly based on outdated info—or outright lies.

5: DEGROWTH

There's a small, though annoyingly loud, group of people who understand all the benefits of nuclear electricity but still think we shouldn't go for it. They believe "degrowth" is the only way to have a sustainable future, that it's not enough to clean up our energy sources—we need to completely change the way we live. We have to make and consume less stuff, thereby using way less energy. Basically, we all need to live like it's 1890, but with FOMO. Whether or not this is even possible without societal collapse is a question currently being debated in a thirty-seven-reply thread between two self-certified economists on X, right below a meme of Karl Marx photoshopped into a Subaru. But in my opinion, degrowth isn't just impractical; it's the wrong way to think about how we use energy.

THE PROBLEM WITH DEGROWTH

In 1890, less than a quarter of U.S. houses had running water—and none had central heating. Sounds miserable, honestly. In 1886, a housewife in California would have to walk almost 150 miles a year simply to get water into her home. And that's just the opening act of her daily misery. The task of carrying in wood or coal, gathering and dumping out ashes, and maintaining a stove took around four hours a day.[1]

In 1945, a study compared steam-powered appliances with electric ones.[2] One brave American housewife dove into washing and ironing a load of laundry the old-school way. It took her four hours to wash and nearly five to iron. Nine hours of her life. On laundry. Then she got to bring in high-tech gadgets—a water heater, electric washer, dryer, and iron. Boom! The same pile of clothes took just forty-one minutes to wash and less than two hours to press. Just imagine how bad things were with metal scrub boards.[3]

There's a popular YouTube account called Early American that recreates daily life activities using only tech from the eighteenth and early nineteenth centuries. In one of its videos, a woman takes her shot at making chocolate ice cream following a recipe from 1830. She mixes all the ingredients in a bowl, then breaks down thick sheets of ice with a hammer. The woman puts the ice in a barrel and proceeds to churn a container of cream by hand . . . for two hours. It boggles the modern mind, but that's what a life with low energy consumption looks like: an endless hellscape of tedious chores to accomplish the most basic things.

Why talk about laundry machines and eighteenth-century ice cream? Because it's important to remember that while human

energy use has skyrocketed over the years, not all of that consumption has been the result of hedonistic waste. A lot of it went into making life *less miserable*.

At its core, degrowth is a critique of capitalism, which is fair. Capitalism, much like fossil fuels, has lifted people out of poverty and improved our access to all sorts of products that can make life easier. However, unchecked capitalism has led to companies cutting corners, overconsumption, and the destruction of natural resources.

My first foray into an American supermarket as a newly arrived immigrant from Brazil was shocking. I spent hours walking through every aisle, utterly amazed by the endless variety of packaged foods. The most mind-blowing discovery was the diversity of breakfast cereal brands and flavors. *So much cereal.* How many different ways can you sell sugary clumps of corn and wheat? The answer is: a lot. Americans can choose from almost five thousand different types of cereal. Are all of them improving lives? Doubtful.

So yes, one could definitely make the argument that capitalism has gone off the rails. But while we patiently wait for a better system to emerge, we should still prepare for the future. And I don't think there's *any* economic system that will make people *want* to use significantly less energy. Just look at what happens when you cross-reference a country's Human Development Index (a simple model for quality of life) and its per capita electricity usage. Quality of life goes up.[4]

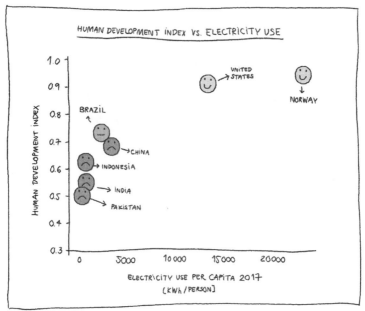

SOURCE: WHAT IS NUCLEAR, 2017

Countries like the United States have more or less leveled out in terms of how much juice they need to meet everyone's basic needs. But plenty of developing nations are still on the come up. Unless you're an actual cartoon villain, you should be able to admit that those countries, and the people in them, deserve to keep harnessing more energy. They deserve to achieve the kind of flourishing that can only happen when your food stays fresh for days at a time, your home is as cool or warm as it needs to be to keep your family comfortable, and there is clean water that doesn't require a pilgrimage to acquire.

I, for one, think it's great that we've engineered our way out of spending four hours a day scrubbing clothes and scavenging for nuts. Now we can spend more time making art, learning something new, or God forbid ... having *fun*. I *like* living in a world where my family in Brazil is just a day's journey away and where I can work and make friends with people from all over.

The concept of degrowth was foreign to me until moving to the United States. That's probably because I grew up in a developing country—where everyone longed for more of the things that make life easier. In my childhood home in southern Brazil, we didn't have luxuries like a laundry machine, air-conditioning, a dishwasher, or central heating.

Some of my most vivid childhood memories involve scrubbing clothes by hand in a water tank during bone-chilling winters. It wasn't just about avoiding a little discomfort; it was about precious time stolen from my childhood—time that could've been spent on anything other than scrubbing dirty socks.

Summer was a different kind of misery. Nights were an endless dance of pillow-flipping, desperately seeking that elusive cool side to soothe my hot head. Only those who've lived without AC know this specific brand of torment.

And that is the point: You can't truly understand what a low energy life *feels* like if you have only experienced it on the occasional glamping trip. So, no offense to folks in developed countries, but you don't get to dictate what the rest of the world should be doing from the top of your climate-controlled ivory tower.

There are other bad outcomes of an obsession with degrowth too. Often environment-conscious humans spend way too much time either beating themselves up or bullying others for being greedy, gas-guzzling monsters. Let me drop a truth bomb. Do you spend a lot of time worrying about your personal carbon footprint? That's exactly what fossil fuel companies want you to do. British Petroleum, one of the largest oil companies in the world, teamed up with a PR firm to popularize the term "carbon footprint" in the early 2000s.

The company even put out a "carbon footprint calculator" in 2004 to help individuals stress over how their daily life choices contribute to climate change. It's one of the stickiest, most successful pieces of propaganda ever. Maybe fossil fuel companies want us focused on our personal choices so it takes the spotlight away from them.

Corporate conspiracies aside, my biggest gripe with the idea that we should ~just use less energy~ is that I'm skeptical any developed, wealthy country would get on board with this plan. Their energy infrastructure is already established, and their citizens would go nuts if forced to give up the comforts of modernity. When American state and local governments started talking about gas stoves, which research shows are *bad for human health* and the environment, some politicians came out swinging. They said that a federal ban on gas stoves would be an affront to liberty, freedom, and the American way.[5] "I'll NEVER give up my gas stove," read a January 2023 tweet from Texas Republican representative

Ronny Jackson. "If the maniacs in the White House come for my stove, they can pry it from my cold dead hands. COME AND TAKE IT!!"[6] This all started because a few state and local officials made moves to keep gas stoves from going into *new* constructions within their jurisdictions. You'd think they were confiscating family heirlooms.

So, yeah. I'm quite confident that the United States and other developed countries aren't going to make drastic cuts in their energy usage. The only logical outcome is that rich societies would keep using as much energy as they want, while poor countries would be prevented from developing because, well, "Sorry, no progress for you. We're saving the planet."

WE WILL NEED A LOT OF ELECTRICITY

Imagine if someone turned on eighteen electric stove burners at full blast and just left them on for the rest of their lives. You'd think they were some kind of wasteful monster (with a weird abundance of stove burners). But that's what pretty much everyone in the United States is doing: The average American uses so much energy that it comes out to a constant hum of 9.5 kilowatts, all day every day.[7] Most countries use less energy than America does, but it still adds up.[8]

People in developing nations will keep increasing their energy consumption, as they should. So how much *more* energy will we need to power the whole world in the future? That's an important answer to get right.

If everyone in the world shared the same luxuries as the average American—reliable artificial lights, refrigeration, AC, running water, transportation, memes—global needs would more than double.[9] And that's with our current population and tech, but things are evolving quickly.

There's a big push across the globe to switch all cars and home heaters to electric in order to get off fossil fuels and cut back on emissions. Tesla CEO Elon Musk projects that to achieve this goal, *electricity** output in America will need to triple compared to 2024.[10]

The truth is future electricity consumption predictions are most likely off. We've officially entered a new chapter in human history: the age of artificial intelligence (AI). Welcome! Please enjoy the apocalypse buffet of large language models, deepfakes, and data centers that consume more energy than bitcoin mining rigs.

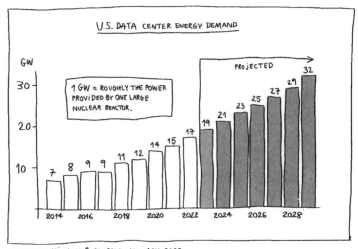

SOURCE: McKINSEY & COMPANY, JANUARY 2023

---

* Energy and electricity consumption are not the same thing. Because newer technologies are more efficient at making electricity, it's possible that we could *increase* our electricity consumption while *decreasing* our total energy consumption.

AI data centers are popping up all over the world and they require exponentially more energy than traditional ones.[11] This new and unexpected demand is leaving utilities scrambling to find ways to fill the gap.[12] McKinsey & Company estimates that by 2029, AI servers in America will require the power equivalent of thirty-two large nuclear reactors.*

The thing is that AI is just one tech revolution adding to our electricity needs. Just imagine how much more energy we will need thanks to innovations we haven't even dreamed of yet.

I'm skeptical of energy demand forecasts beyond ten years. It's like someone in 2005 trying to predict how much energy we'd use in 2025. No one back then could have foreseen our absolute devotion to smartphones. I'm also confident that not one "energy expert" could have predicted that people would devote entire warehouses full of machines to mining imaginary coins with dog faces. In other words, we can't be certain exactly how much energy we'll use in the future. But one thing we can be sure of is we're not going to use any less of it.

Global energy consumption has gone up almost every year for more than half a century. The exceptions to this are not encouraging. It went down in the early 1980s, due to the oil crisis in the 1970s. It also dropped in 2009, following the financial crisis, and in 2020 during the COVID-19 pandemic.[13]

Want to be depressed for a minute? We've accidentally already run a mini experiment on degrowth. In 2020, the world virtually came to a halt during the COVID-19 pandemic. Entire cities shut down for months with hardly any gas-guzzling cars on the roads

---

* While writing this book, the predictions for AI's electricity demand were rapidly evolving as new studies and technologies developed. As I said before, I believe these forecasts are usually dumb.

and no fuel-hungry planes flying through the skies. Even though the production of goods and services was drastically reduced, CO_2 emissions fell by a measly 5.4 percent that year.[14] People still needed a lot of energy to survive. A study from NASA noted that emissions bounced back to near pre-pandemic levels by the end of that same year.[15] The only way to reduce emissions permanently is to transition to low-carbon-emitting technology.

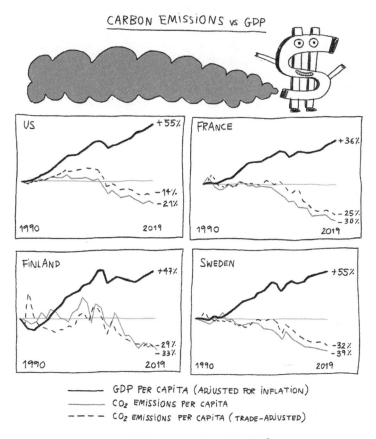

SOURCE: HANNAH RITCHIE, NOT THE END OF THE WORLD, 2024

If your master plan for saving the planet involves everyone giving up electricity and becoming off-grid forest witches, you're wrong. We now have plenty of evidence that economic growth doesn't *have* to lead to more carbon emissions and environmental destruction. In places like Sweden, France, and the United States, carbon emissions have gone down even as economies continue to grow.*

Instead of focusing on cutting back, we should prioritize building out the tech we already have.

Because the problem isn't that we use energy. The problem is where it's coming from.

---

* Yes, this also accounts for the fact that some industries might have moved overseas.

TL;DR

Degrowthers drive me nuts. In my experience, they almost always come from energy-rich societies and yet somehow romanticize poverty, like it's some kind of spiritual awakening. Contrary to what proponents of this philosophy may tell you, using lots of energy isn't inherently bad. In fact, energy abundance is the secret to human happiness and equality. We've already figured out how to grow economies without increasing emissions. All we need to do is sub out the dirty fuels of the past with clean ones. It's not rocket science, unless you count nuclear as rocket science, which, let's be real, it kind of is.

6: CLEAN ENERGY REVOLUTION

The only future worth fighting for is one of abundance for all, where our energy sources have the least amount of impact on the environment and humans. It's time for a clean energy revolution. Let's explore the key sources of clean energy, how they work, and how we can make better choices.

100 PERCENT RENEWABLE

You might have heard that we need to replace fossil fuels with "100 percent renewable energy." But there's a problem: Nuclear electricity, while clean, isn't *technically* renewable. More on that later, but for now let's take a look at the origins of the renewable energy craze.

The slogan can be traced to a paper in *Science* published in 1975 by Danish physicist Bent Sørensen.[1] But it was American energy nerd Amory Lovins who popularized the idea. In the 1970s,

Lovins was a passionate representative for the environmental organization Friends of the Earth. Friends of the Earth has a curious backstory.

It started in the late 1960s with David Brower, a renowned mountaineer and conservationist. Brower had spent over two decades working with the Sierra Club, one of the most prominent environmental organizations at the time. But a dramatic fallout was brewing. Brower was against the Sierra Club's pronuclear electricity stance, a beef that ultimately led to his dramatic departure.

Effectively, Friends of the Earth was founded as an antinuclear environmental organization. The founding donation of $200,000 was provided by Robert Orville Anderson, the owner of—drumroll, please—the Atlantic Richfield oil company.[2] Yes, an oil magnate jump-started an environmental group dedicated to fighting nuclear electricity. Can't say I'm shocked.

I'm not trying to accuse Brower or Friends of the Earth of shilling for big oil. But it's interesting that a fossil fuel mogul would write a big check to a brand-new organization. It's especially interesting, considering that Friends of the Earth's main difference from far more established environmental nonprofits was that it was advocating against nuclear.

Amory Lovins later pivoted from nonprofit work into full-blown energy policy influencer, and his ideas took off following the 1973 oil shock. This is when petroleum-rich nations in the Middle East decided to place an embargo on oil exports to certain countries, including the United States. Oil prices went through the roof and energy independence became a hot topic. Lovins became a prominent advocate for renewable energy and a vocal opponent of nuclear, as evidenced by his book *Non-Nuclear Futures: The Case for an Ethical Energy Strategy*. He is basically the daddy of the "100 percent renewables" idea.

Lovins and his modern disciples believe we can get all the energy we consume from "renewables" like solar, wind, and hydro. Very sunshine-and-rainbows. But, as always, the devil is in the details.

Nuclear isn't technically considered a renewable energy source. Even though uranium and other nuclear fuels are abundant in nature, they are finite.

The term "renewable" refers to energy that can be replenished or doesn't run out. Think solar, wind, hydro, or geothermal. But under this definition, wood, food scraps, and even cow poop is considered renewable. Yes, literal poop. Burning these things creates pollution and releases greenhouse gases, both of which we're trying to avoid.

"Clean" or "zero-carbon" energy, on the other hand, means energy sources that produce little to no greenhouse gas emissions or air pollution. The shocking truth is that 100 percent renewable does *not* mean 100 percent clean, even though these terms are

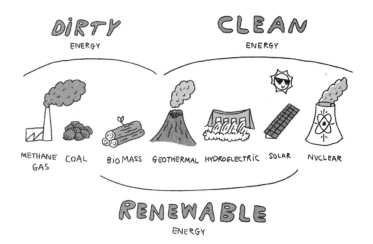

often used interchangeably. It's just a sneaky rhetorical strategy used to exclude nuclear electricity.

And there's another problem. While the sun and the wind *are* renewable, solar panels and wind turbines are *very much not*. They are made of materials like silica, concrete, aluminum, steel, and rare minerals, which are all *finite*. They can last for a decent amount of time—around twenty years for turbines and up to thirty for solar panels—but that's very far from forever. More upsetting is the fact that chopping down trees and burning their wood to make electricity counts as "renewable energy."

A couple of decades ago, Lovins and his groupies arbitrarily decided we *must* get to 100 percent renewable energy. Since then, every climate conference, political speech, and anti–fossil fuel rally has reliably featured "100 percent renewable energy" signs and chants. While it caught on, the obsession with this purity-tested goal just makes it harder to build a clean, reliable, and abundant energy future. And let's be honest: the only reason this 100-percent renewable thing stuck around is because people are still scared of nuclear.

Since the 2000s, several papers have been published saying it is *technically* possible to power the United States using only solar, wind, hydroelectric dams, and batteries.

Getting into a debate on whether it is *technically* possible to do it all with renewables is a waste of time. It is *technically* possible to meet your daily caloric needs by only eating cauliflower. Sure, you would need to eat about 20 pounds of cauliflower a day. Which means taking several trips to the supermarket, having a bunch of fridges to store all of it, and spending a big chunk of your day chewing. In other words, it would be extremely impractical. Why go through all that trouble when there are other delicious and healthy food options? It's also *technically* possible to walk 2,900 miles from San Francisco to New York City. But that's a dumb thing to do, considering airplanes exist.

Our goal shouldn't be to rely solely on "renewable" energy sources—that's silly. The sooner we stop playing semantic games and start prioritizing clean energy (*all* clean energy), the faster we get to a future that's both sustainable *and* livable. And here's a crazy idea: Maybe it's time to stop calling burning trees a form of renewable energy.

CLEAN ENERGY

We already talked about my favorite source of clean energy: nuclear. But there are plenty of others to choose from, each with its own strengths and challenges. No single source can meet all of our needs on its own—but the good news is, it doesn't have to. By mixing and matching, we can create energy systems tailored to the resources available in each region. Let's take a closer look at some of the other options we get to work with.*

SOLAR

We've been capturing energy from the sun since forever. In ancient Rome, architects built bathhouses with large south-facing windows that were glazed to retain heat. Of course, that's *very* different from our modern tech, but the science behind our modern use of solar power is shockingly old too. You might be surprised

---

* I can already feel the promoters of some obscure clean energy source seething right now. I am not mentioning every single source out there, just the main ones.

to learn that the first solar panel* was created in 1881.[3] With an efficiency of less than 1 percent, it managed to convert only a tiny fraction of sunlight into electricity. Newer technologies are a lot better than early models, with some high-end ones like thin-film solar cells reaching efficiencies of up to 29.1 percent. What's more impressive is the cost of solar panels has nose-dived by as much as 89 percent in ten years, making solar one of the cheapest sources of energy and a lot more accessible.

In 2023, solar generated 5.5 percent of the world's electricity. That's more than ten times its share of the global grid just a decade earlier.[4] The United States got 3.9 percent of its electricity from solar in that same year.[5]

Despite the sun's abundant energy, when it decides to hide behind clouds or set for the night, our solar tech can't produce power. That means no city can run exclusively on solar without some kind of backup.

WIND

Sails, which date back to at least 3300 BC, represent humanity's first success in harnessing breeze. But sailing is just one way to catch the wind. The Persians started spinning up windmills sometime around the ninth century, and later the tech spread to Europe, landing in the Netherlands around the fifteenth century. These early wind-powered machines were used to grind grain, pump water, and saw wood.

---

* Solar panels work by capturing sunlight with cells made of materials like silicon, which have a special property: They can absorb photons from sunlight, kicking electrons loose. This movement of electrons creates an electric current. When sunlight hits the panel, it's like the photons are playing a game of tag with the electrons, sending them zipping through the panel's circuitry. This current is then gathered and sent along wires to our homes and businesses, lighting up our lives.

Folks figured out how to make electricity with wind power in the 1880s. They took the grain-grinding windmills from ancient history and stuck a bunch of magnets inside.* Similar to solar panels, the cost of wind turbines has dropped by as much as 70 percent in ten years. In recent years, costs have fluctuated though, in some cases going up by 25 percent.

In 2023, wind amounted to around 10 percent of electricity in the United States and 7.33 percent of electricity generation worldwide.[6] Wind's slice of the global electricity pie has doubled since 2015.

I love that we can capture the wind and use its energy to charge our phones. But, just like the sun, the wind sometimes goes AWOL and wind turbines just stand still, waiting for the next gust of air.

HYDROPOWER

Water isn't just a must for dewy skin; we've also been grabbing useful energy from H_2O since the BCs. Chinese engineers started working on waterwheels on rivers around the fifth century BC, and they were using more sophisticated models to mill grain by the first century AD. Meanwhile—sometime around the third century BC or so—ancient Greeks were independently inventing their own waterwheels. They used them for moving water, milling grain, and eventually mining. There's some evidence that they even stuck waterwheels onto boats to serve as super-analog odometers. But did they stress about getting their steps in? We'll never know. To turn hydropower into hydroelectricity, we just use the same magical wire-and-magnet combo that powers wind tur-

---

* Wind turbines have big blades that spin when the wind blows. This spinning motion turns a generator packed with wires inside a ring of magnets, which interact to make electricity. The electricity is then sent through wires to homes and businesses.

bines. Water flows, spinning turbines that are connected to electric generators, and that's it.

Hydropower was responsible for 5.7 percent of all electricity in the United States in 2023.[7] Globally, it provided 15 percent of all electricity that same year.[8]

Throughout the twentieth century, hydroelectric dams expanded quickly and today they're the largest source of clean energy in the world. Though hydropower is more reliable than solar and wind, it's still dependent on geography and the weather. Dams can only be built on large bodies of water with a lot of oomph, and during drier seasons their output can go down dramatically.

GEOTHERMAL

If you've had the pleasure of soaking in a hot spring, you've personally witnessed the power of geothermal energy. In fact, our deeply ingrained love of soaking and shvitzing might be older than our species. Japanese macaques, aka snow monkeys, are known to dip into hot spring pools during cold winter days. Scientists have shown that the spa treatment lowers their stress levels. Same.

Rocks closer to Earth's surface have radioactive elements like uranium, thorium, and potassium. These elements naturally break down, or decay, over time, releasing heat. Geothermal plants tap into this subterranean heat by drilling wells deep into Earth's crust to reach hot rock layers. Depending on the location, this might be anywhere from a few hundred to several thousand meters deep.*

---

* To make electricity, cold water is pumped down into these hot rocks through an injection well. The high temperatures underground heat the water, turning it into steam or very hot water. Finally—you guessed it—that steam spins a turbine connected to a generator.

In 2023, geothermal power plants produced around 0.4 percent of the world's total electricity. In the United States, that number was the same.⁹

Geothermal resources are not uniformly distributed across Earth. They are more accessible in areas with tectonic activity like the Ring of Fire around the Pacific Ocean. This geographical constraint limits where it makes economic sense to build geothermal plants.

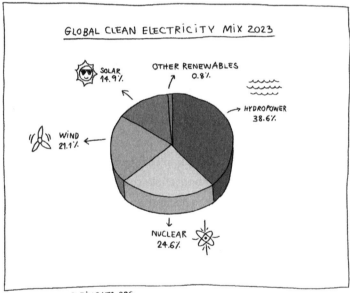

SOURCE: OUR WORLD IN DATA.ORG

THE ULTIMATE ENERGY DIET

Nutrition is a spicy topic. Nearly everyone on the internet has an opinion about what others should eat. But if you cut through all the influencer-sponsored content and chronically online liver-eating dudes, you'll find that most experts say something painfully unsexy: Just eat a variety of healthy foods.

The same thing is true when it comes to energy sources: Tune out all the yelling from folks pushing their one true energy gospel and you'll see what we really need is a full food pyramid. I like to call this the "energy diet."

The energy diet is a really simple concept: Countries should be making the most of their local resources and getting their energy from a variety of options—ones that don't kill people or destroy the planet. Sunny city? Put solar panels on every roof. Windy area? It's turbine time. Sitting on a geothermal jackpot? Drill, baby, drill (but in a green, better-for-the-planet way). Big bodies of water with a lot of power? Hydropower dams. And if you don't have any of those things or enough land, but still want clean energy? Nuclear! Easy, right? Except we know it's not. If it were simple, we'd all be savoring mouthfuls of clean, delicious, healthy energy by now.

In this nutrition metaphor, fossil fuels are the ultra-processed junk food. They're fast, convenient, and cheap . . . basically the gas station taquito of the energy world. And just like junk food, we're addicted. It's what most of us grew up on, it powers the economy, and it's everywhere. But we also *know* that an abundance of fried foods and preservatives is bad for us.

Just as it is when it comes to accessing whole and nutritious foods, there are parts of the world and socioeconomic groups where it's practically (or literally) impossible to sub in cleaner fuel options. A person who lives in a food desert and works a stressful job isn't evil or ignorant for grabbing something from a drive-through to make sure their kids get fed. But we can all agree that the world would be better off if it were just as easy for them to get a fresh and wholesome meal instead.

Technically speaking, fossil fuels like coal and oil have helped lift people and countries out of poverty—just as cheap, unhealthy food has no doubt kept countless children from starving. But that doesn't mean we can't do better. It's easy to conflate energy with fossil fuels, because they have been our predominant source of energy for so long. Access to energy is what lifts people out of poverty, not fossil fuels themselves.

Let's zoom out to look at the rest of the energy shopping cart.

Solar and wind are definitely an upgrade over a bucket of fries. They're like the celery and cucumber in our energy diet: healthy and important, but tough to fill up on. Because of the intermittent nature of the sun and the wind, you can't rely on them 24/7. It's like only being able to eat from a single restaurant with finicky hours. When the owner decides to close up shop and take a nap, your plate is left empty.

Geothermal is like one of those sugarloaf pineapples you can only find in Hawaii: nutritious, goes down easy, and packs more calories than celery, but not always easy to come by. Geothermal wells are more reliable than solar and wind, but you can't just plant one anywhere and expect it to work its magic. It's a bit like having a specific soil requirement for those unique pineapples—it needs the right geological conditions. Geothermal wells tap into Earth's natural warmth, which is a lot easier when features like volcanoes bring the heat closer to the surface.

Hydropower is our rice. It's delicious and nutritious, and fairly easy to grow . . . as long as you have *plenty* of water. Hydroelectric dams are at their best when they're located in areas with abundant rivers and rainfall, turning that water flow into a steady stream of power. Hydropower's efficiency is closely tied to the reliability of water flow, making it a bit like depending on rice fields in regions with unpredictable rainfall.

Nuclear is the peanut butter of energy—sans allergy. Making it is a little more complicated and expensive than just picking a vegetable and putting it in a box, but it packs a ton of calories in a small portion. It's also healthy and delicious, if you don't go throwing in a bunch of ingredients you don't need. Nuclear's superpower is its incredible energy density. It's like having a jar of peanut butter that lasts for years. Nuclear's ability to produce large quantities of electricity consistently makes it a powerhouse in the energy buffet, offering a reliable option to satisfy our ever-growing appetites.

Batteries play the role of coolers in our energy landscape—they don't create electricity but are great at storing it, holding on to what's produced from different sources until it's needed. Imagine preparing for a picnic: An empty cooler will leave you hangry, and a poorly packed one will leave you filled with regret. However, a well-prepared cooler stocked with fresh veggies and peanut butter sandwiches lets you keep the party going without the crash and burn.

A good cooler will keep piles of veggies fresh, but you'd need an obscenely large one to store enough celery and cucumbers to meet all your caloric needs for the week. We run into the same issue with solar and wind.

Some folks say that batteries would make solar and wind reliable enough to run the whole grid. But it's a little more complicated than that. Day after day, clouds rudely float in front of the sun, reducing solar panels' capacity by as much as 70 percent.[10] That's when lithium-ion batteries can save the day: They release stored electricity into the grid and prevent society from collapsing until the sun comes out again. But sometimes the sun barely shines, or the wind barely blows for *days* at a time. The Germans even have a name for it, because of course they do: *dunkelflaute*, which roughly translates to "dark doldrums." These lulls create a massive challenge for a grid reliant on renewables only. In November of 2024, low wind speeds left German wind farms generating barely 7 percent of their listed capacity. At the same time, wind farms in the United Kingdom only met 3 to 4 percent of demand.[11] While renewable energy sources have some level of predictability, variability is still a problem. Year-to-year changes can be huge. For example, in May 2023, Texas saw a 40 percent drop in wind power generation compared to the same period in the previous year.[12]

Most of the lithium-ion batteries available today have a maximum storage capacity of four hours.[13] It would take a buttload of them to prop up the grid during those times. There have been some recent advances in long-duration battery tech, but researchers say we have a ways to go.[14]

In a 2021 analysis, a Stanford professor claimed the state of Texas could get to 100 percent clean energy with wind, hydro, solar, and batteries only.[15] He claimed long-duration batteries weren't necessary—we could rely on the ones that already exist. The only problem is that it would take 13.4 terawatt-hours of lithium-ion batteries to back the grid up—which is an insane thing to propose when, at the time of that analysis, the *entire world* made 1.57 terawatt-hours of lithium-ion batteries *in a year*.[16] Meaning, we'd need to dedicate our entire planet's battery-manufacturing infrastructure solely to the task of cleaning up the *Texas* energy grid for almost a decade to make this happen. Unhinged.

Pair those panels, turbines, and short-duration batteries with nuclear reactors, and you've got yourself a stable and reliable grid. Nuclear adds a steady flow of electricity into the power lines while solar, wind, and batteries do their elaborate dance throughout the day. Everyone is happy.

To understand why it's so important to maintain a balanced energy diet, just take a look at Sweden, where the tea on nuclear electricity policy is hot.

Back around the 1970s, Sweden had nuclear reactors popping up almost as fast as you could build an IKEA bookshelf. But when the Three Mile Island* accident happened in 1979, Swedes started having second thoughts and slowed their roll. They decided to shut down all their nuclear reactors by 2010.[17] Perfectly good nuclear plants were closed before their time. But the government had to face the facts: Their energy needs were growing because the country was thriving, and that meant they needed more always-on clean power.

---

* We dive deeper into the Three Mile Island accident on page 146.

When 2010 came around, the Swedish government decided that some nuclear plants could stay—and those could even be replaced with new reactors when the time came for an upgrade. Fast-forward to 2024, with Europe in an energy crisis, thanks partly to the war in Ukraine, and Sweden is suddenly looking to have a nuclear renaissance. The government is putting money into research, planning to build new reactors, and even changing targets from "100 percent renewable" to "100 percent fossil free" electricity by 2040—which is what every country should be aiming at.

Fortunately, Sweden came to its senses while it still had six working reactors. Even when public opinion on nuclear electricity was at its lowest, the country still got a huge chunk of electricity from it. That's given Sweden one of the cleanest electrical grids in Europe. What's more, they also got citizens to switch their oil-powered home heaters to electric ones. As a result, greenhouse gas emissions from heating have dropped by 95 percent since 1990.[18] The government managed to get people to make the switch by introducing a carbon tax, which made the price of oil more expensive. But what really got people switching over was the fact that electricity was very cheap because of the country's nuclear power plants and hydropower dams.

As great as this little comeback story is, just imagine how much Sweden might have accomplished if it had stayed full speed ahead on the nuclear train for all that time. This was Sweden's electricity plate in 2023:

Clean Energy Revolution

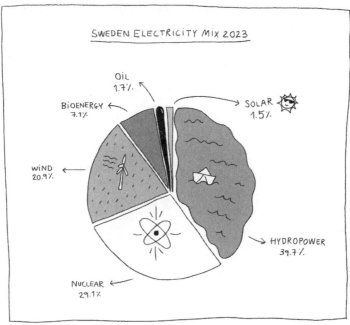

SOURCE: OURWORLDINDATA.ORG

Not bad, considering this was what the world's electricity plate looked like in 2023:

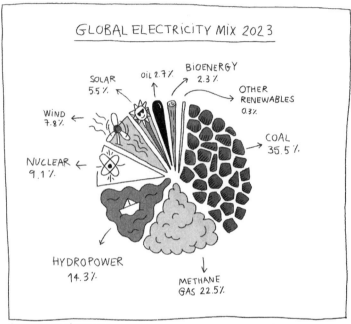

This is what the United States' electricity plate looked like in that same year:

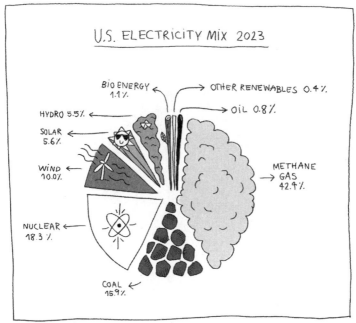

Considering what we know about fossil fuels, neither of those last two plates seem great. But there is hope. Sweden isn't the only country that's cleaned up its plate.

Look at Iceland:

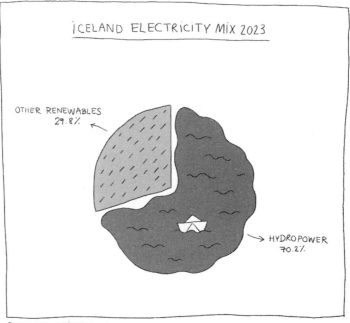

Or Switzerland:

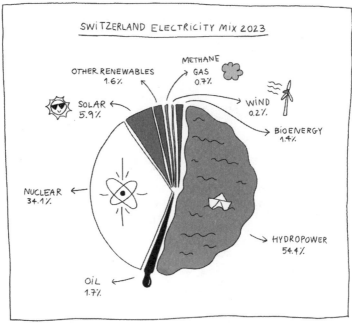

Or Ontario in Canada, as we saw in chapter 3:

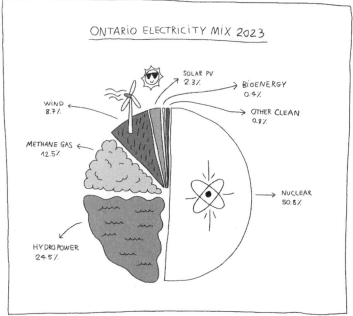

SOURCE: ONTARIO ENERGY BOARD

Most of the places that have cut junk electricity out of their diets look similar. They primarily rely on hydropower or nuclear electricity. It's not that our celery isn't important; in sunny areas, putting solar panels on a house can provide enough power to handle air-conditioning, lighting, and other household needs. That's amazing! We love celery.

It's just not going to be practical to put up enough panels or turbines to keep everything going, everywhere, all the time—and we'd need a breakthrough in battery tech to have the money, materials, and space to fix all those problems by storing wind and solar energy for later. Why go through that hassle if we can complement with other options?

We definitely have room for more rice (hydropower) in our diet too. But there's only so much water in the world. Even when experts imagine a future where we tap every potential hydro source that isn't a state park or otherwise beloved scenic waterway, the result is merely a dent in our reliance on coal and other fossil fuels—not a one-and-done replacement.[19]

What *should* the world's electricity plate look like? Well, it depends. In a perfect scenario, each country would use the resources available to them. Some would get most of their electricity from hydropower, others from wind, or maybe nuclear. At the end of the day, all that matters is that it's coming from carbon-free sources.

What's cool about nuclear is that it's not dependent on geography. You can build reactors that run all the time, pretty much anywhere. So, they can lend a hand whenever locations have tapped out their other energy sources.

More nuclear electricity is the key to a balanced energy diet. And it's not just a badass, reliable energy source. It's also one of the safest and cleanest ways of making electricity. So, now you might be asking yourself: How did we miss the mark on this technology in such a major way?

In the beginning of the book, I said the reason why we turned away from nuclear electricity was complicated and fascinating. There's no denying that its original sin was the bomb, but that's not the full story. Public perception has changed dramatically throughout the decades, for reasons that had very little to do with nuclear electricity itself. Sit back and strap in. We're about to go on a wild ride throughout history to try to understand where it all went wrong.

TL;DR

The quest for clean energy is like trying to create the perfect diet for the planet. We've got solar and wind as our veggies, geothermal and hydro as our carbs, and nuclear—the peanut butter that packs a punch with minimal environmental guilt. We've been chasing the dream of 100 percent renewables like it's the latest diet fad, but here's the tea: "Renewable" doesn't always mean "clean." Burning wood for electricity is like saying you're on a health kick while vaping. Countries should play to their strengths: If you've got sun, go big on solar; if you've got wind, let those turbines twirl. And if you can go nuclear, do it. It's reliable, efficient, and balances everything out. The world's current energy plate needs a revamp, less fossil fuel junk food, and more of this balanced clean energy diet.

PART 3

DIPLOMACY: FIGHTING A WORLD AGAINST NUCLEAR

7: CHANGING TIDES

NUCLEAR BEFORE NUKES

When most people think about the history of nuclear electricity, their timeline starts somewhere around the Manhattan Project in 1942. But before humanity reckoned with the atrocities of World War II or spent decades running duck and cover drills thanks to the Cold War, our kind had a very different view of radiation.

In his book *The Rise of Nuclear Fear*, historian Spencer R. Weart explains that when scientists first began to explore radium at the start of the twentieth century, they spoke about it as if it were the key to fairy-tale alchemy—the transmutation of air into gold.[1] Not *precisely*, but close enough for the general public to get extremely hyped.

One important thing to consider when talking about this early twentieth-century atomic fervor is that no one knew *shit* about atomic energy. Sure, scientists were making significant discoveries, and progress was happening in real time. But physicists knew very little about atoms, fission, fusion, radiation, or any of the rest of it.

So, when folks waxed poetic on all the possible applications of atomic energy, they weren't *really* talking about atomic energy. They were using atomic discoveries as the brush with which to paint a picture of the sci-fi utopia they'd always dreamed of.

Now, as it so happens, they weren't too far off the mark. Okay, they may have gone a little wild in comparing atomic energy to *actual* fairy-tale magic, but they were right in supposing that atoms could power our world and make inequality a thing of the past. That's exactly the future I'm hoping we can still fight for.

If folks in the early 1900s had these same rad future dreams, then why don't we live in a nuclear-powered wonderland today? The short and 2D answer is easy, and we've already covered it: the bombs. But some historians, like Weart, argue that our outsize fear of nuclear rose *precisely* because we once had such an optimistic view of it.

In the early twentieth century, radium wasn't just a scientific curiosity; it was a magical substance that seemed to promise health, vitality, and even immortality. That extended beyond the world of energy and into the general public's medicine cabinet. Snake oil salesmen peddled glowing potions and radioactive elixirs, promising everything from a cure for cancer to a solution for low libidos. From tonics and salves to suppositories (yes, really), the market was flooded with an array of radioactive concoctions, each claiming to possess extraordinary healing properties. One infamous example was the Radium Ore Revigator, a water crock lined with radioactive ore. Marketed as a health tonic, the jar was supposed to be filled nightly and users were encouraged to consume several glasses of the radium-infused water daily. It promised vitality by delivering a potent dose of radiation, with claims that it could treat ailments like arthritis and flatulence. Instead, it unknowingly put the health of those who partook at risk. One socialite and golfer (you can't make this stuff up) named Eben Byers actually died from radium-induced cancers because of all the irradiated water he guzzled to amp up his performance on the green.[2]

Like we talked about, low doses of radiation aren't nearly as dangerous as most people think. But that doesn't mean you want to *drink water out of a radioactive pot* manufactured with zero research or regulatory oversight.

The world's willingness to embrace radium as a wonder substance before science had time to catch up made hapless workers sick too. In the 1910s and 1920s, radium-based paint became wildly popular for coloring everything from house slippers to clockfaces so that they'd glow in the dark. But all that flashy paint had a dark side: It poisoned the now-infamous "radium girls."

Girls working in clock and watch factories used a technique called "lip-pointing" to sharpen their paintbrushes. They stuck the dirty brushes into their mouths to create fine points and execute tiny details in their paintings. Radiation from the paint accumulated in their bones and made them sick. Many experienced a horrifying symptom known as "radium jaw" where the bones of their faces literally rotted away.

Companies that made money off hawking radium did everything they could to discredit the girls, including claiming they'd died of syphilis. By the time the truth came to light, the public was rightfully freaked out about the whole situation. No one likes being lied to, and no one wants to find out their favorite watch helped cause the painful death of innocent young girls.

But the story of the radium girls isn't about radiation at all. The people producing the paint knew it was dangerous to ingest. Wearing a watch with a bit of radium paint in it is perfectly safe, but those girls should have been wearing masks and gloves. There are a lot of things inside a watch that I wouldn't recommend licking, tbh. There's a similar condition to radium jaw called "phossy jaw" that girls who worked in matchstick factories suffered from, caused by white phosphorus. That stuff is super toxic: Inhaling fumes for a few years in a row was enough to make people sick. But white phosphorus isn't radioactive, and for some reason we hear about the phossy jaw girls a whole lot less. And we won't even get into the countless people who've gotten sick and died from mining coal.

Meanwhile, people writing for magazines like *Popular Science* continued to speculate wildly on what universal secrets atomic science might unlock—and started imagining all the ways nuclear physics might run amok. There were death rays, earth-exploding nuclear chain reactions, and mutated creatures that experienced millions of years of evolution in the blink of an eye. Then comic book and sci-fi movie writers took those ideas and ran with them faster than a subatomic particle.

By the time the U.S. government started working on building the first atomic bomb, people's imaginations about nuclear's potential for both creation and destruction were running wild. Lots of scientists worried about how all the secrecy around the Manhattan Project might come back to bite us. They knew that nuclear electricity had loads of potential, but they also knew that nuclear bombs would terrify people. They worried the public would get the two confused. Some scientists anticipated this before the bombs dropped, discussing what kinds of educational materials they could release once the war ended.

But alas, the bombs dropped in 1945, and all the hopeful visions of atomic science were overshadowed by its use as a weapon of mass destruction. Atomic research became a symbol of all the ways scientists and politicians had overpromised on their ability to change the world—and all the things they'd secretly mucked around in.

ATOMIC FERVOR

For the first *fifteen* years after the discovery of nuclear fission, reactor designs were classified—top secret, James Bond–type stuff.

They could only be built *by* governments *for* governments. Meaning, they were used exclusively for research, making bombs,* or powering submarines. It's no surprise, then, that the public began associating nuclear with the military and bad juju, rather than seeing it as a potential force for progress or peace.

As a way to get everyone to chill after World War II, American president Dwight Eisenhower delivered a speech at the UN General Assembly in 1953. It was called "Atoms for Peace." The speech was the launch of a PR campaign with the goal of showing the rest of the world that the United States was more interested in peace than in war. The hope was to calm down fears of continuing nuclear weapons armament by promoting the peaceful uses of nuclear instead, like electricity production, medicine (like CT scans or radiation therapy for cancers and diagnostics), industry, and agriculture.†

One of the campaign's successes was the creation of the International Atomic Energy Agency in 1957. The IAEA is basically the referee of nuclear activities around the world. It promotes the peaceful use of nuclear technologies, while slowing down their use for military purposes.

In 1970, the Treaty on the Non-Proliferation of Nuclear Weapons (or NPT for short) went into effect. The NPT is like a global restraining order against proliferation. Almost every country has signed on, with notable exceptions like India, Pakistan, Israel, and North Korea, which signed, then rage-quit later. The treaty has three main objectives: Stop the proliferation of nuclear weapons, promote cooperation in the peaceful use of nuclear technology, and commit to

---

* Some types of nuclear reactors, not all, can be used to produce plutonium-239 as a byproduct. Plutonium-239 is a key material for making nuclear weapons.

† Nuclear in industry is great for inspecting and sterilizing stuff without ruining it, while in farming, it's all about making super crops, zapping pests, and making sure our food lasts longer and our soil behaves.

disarmament. Despite criticisms, especially concerning the disarmament part, the NPT has played an important role in curbing the spread of nuclear weapons. The IAEA has been responsible for enforcing the NPT's safeguards since its inception. This includes doing inspections at nuclear facilities globally to ensure countries are only using those facilities for the good stuff.

Another success of the Atoms for Peace campaign was that it got people psyched about nuclear electricity. It became a symbol of freedom, prosperity, and scientific innovation. This newfound excitement could be seen everywhere, from hyper-futuristic architectural monuments to nuclear-powered cars (terrible idea, 0/10 do not recommend). Designers started dreaming what our rad future could look like. You can see this vision clearly manifested in the Atomium, an iconic stainless-steel structure in Belgium that was built in 1958. It's a 335-foot-tall sculpture that was intended to showcase Belgium's postwar technological and scientific prowess, with a focus on the peaceful use of atomic energy.

1957 Ford Nucleon concept car

Atomium in Brussels, Belgium

Even American dream-weaver and mouse overlord Walt Disney played a role in renewing this love for atomic science. The episode "Our Friend the Atom," which premiered in 1957 as part of the *Disneyland* TV show, aimed to educate the public about nuclear and its applications for good. The man, the myth, the legend, Disney himself explains how nuclear fission works and can be used for many things (other than bombs). Fun fact: Walt Disney thought nuclear electricity was the future and wanted his amusement park to be powered by it. To this day, Disney World in Florida has a license to build and operate a nuclear power plant. So if someone out there wants to make dreams come true *and* decarbonize the grid, you know what to do.

Nuclear was again marketed as the miracle fix, the thing that would finally lift humans out of darkness and send us straight into our *Jetsons*-like, God-given future. In 1957, almost twenty years after the discovery of nuclear fission, the Shippingport Atomic Power Station was inaugurated. It was the first *commercial* nuclear plant in the United States, and the first in the world to be built *exclusively* for the purpose of generating electricity for the public. But it wasn't until the 1960s that things really took off. As women were sharpening their cat eyes and marching for civil rights, countries went all in on nuclear electricity, with seventy-four reactors popping up worldwide from 1960 to 1970.

Still, not everyone was sipping the atomic Kool-Aid. Skepticism lingered because of the initial secrecy around nuclear research. Many thought the Atoms for Peace campaign was icky and just a cover-up by the American government to keep stockpiling nuclear bombs. There was a sense that nuclear electricity was just another Big Government scheme. This vibe would end up setting the stage for the rise of the antinuclear movement.

MIXED FEELINGS

While reactors were springing up in the 1960s, society was also approaching peak Cold War anxiety. The United States and the Soviet Union had been stockpiling and testing nuclear weapons since the beginning of the atomic age in the most obnoxious dick-measuring contest ever. Generally speaking, these tests were conducted in "uninhabited" areas that we now know were very much inhabited by Indigenous people, people of color, poor rural communities . . . You get the picture.

I could spend a few dozen pages talking about how horrifying and inhumane all of this was, but I'll stick to just one example that puts it all into perspective. In the 1940s and 1950s, the United States conducted sixty-seven nuclear weapons tests on two of the Marshall Islands' northern atolls, Bikini and Enewetak—idyllic, tropical paradises in the Pacific Ocean with crystal clear lagoons, lush palm trees, and vibrant marine ecosystems. In preparation for testing, the people of Bikini were relocated to other islands. Some Bikinians were permitted to return home in the early 1970s after assurances of safety. However, by 1978, they were evacuated *again* due to high radiation levels detected in their bodies. Seven decades after the first tests, in 2019, researchers from Columbia University tested several islands and atolls for radioactivity. They found that certain places have ten to one thousand times more radiation than areas around Chernobyl or Fukushima. The legacy of displacement and contamination left deep scars on the Marshallese people and was just one chapter in the growing unease around nuclear technologies during the mid-twentieth century.

As the world grappled with the consequences of atomic blasts, tensions escalated between superpowers, bringing humanity dangerously close to Armageddon. At no time was this more evident than during the Cuban Missile Crisis.

On October 22, 1962, President John F. Kennedy sat at his desk looking calm, cool, and collected as he told the American people that they were, in fact, maybe about to die. Earlier that month, an American spy plane had gotten evidence of Soviet nuclear missile sites being built in Cuba. The United States responded by creating a blockade around the island country. In the days that followed, there were loud public threats of a nuclear exchange. For an entire week, Americans went to bed wondering if they'd wake up at all, while the rest of the world braced for the possibility of full-scale nuclear war. Global annihilation felt like a button-press away. On October 28, the Soviet Union announced they would dismantle and remove the missiles from Cuba, and tensions eased up a bit.

The Cuban Missile Crisis sent a very clear message: *Maybe let's not blow up the planet.* That urgency, combined with growing fears of contamination from aboveground nuclear weapons testing, led to the Partial Test Ban Treaty in 1963. The treaty, signed by the United States, the Soviet Union, and the United Kingdom, banned nuclear testing, including for peaceful purposes, in the atmosphere, under water, and in space.

It wasn't enough. People demanded a complete halt to nuclear weapons testing—and who could blame them? Decades of radiation exposure, wars, weapons testing, and countless shady government cover-ups had eroded public trust. By then, it was nearly impossible for people to separate their deep distrust of all nuclear technologies from the promises of nuclear electricity.

THE ANTINUCLEAR MOVEMENT

While concerns for the environment had existed for centuries, it wasn't until the 1960s that a more formal movement started to take shape. In 1962, a book called *Silent Spring* by marine biologist and conservationist Rachel Carson caused a stir in America. It revealed how chemicals like DDT were wreaking havoc on birds, bees, and essentially all of nature.

This was a time when many companies simply dumped their

chemical waste into rivers, lakes, oceans, and open landfills without much oversight, and it obviously sucked. To name just one example, the Cuyahoga River in Ohio caught fire *more than a dozen times* throughout the 1960s. The fires were caused by an intense concentration of oil and other human-made nastiness.

As people became increasingly aware of environmental issues, the broader social justice wave of the 1960s and 1970s was brewing. Movements like the anti–Vietnam War campaign, civil rights activism, women's liberation, the student movement, and the rise of counterculture inspired mass protests and widespread calls for change. Supporters of the counterculture movement rejected authority and traditional structures of government, seeing them as corrupt or oppressive. There was a healthy distrust of The Man. Remember, they were emerging from an era filled with government deception and overreach, so when it said, "Hey, trust us with nuclear," people understandably went, "Yeah, no thanks." In many ways, the environmental movement was born, and grew up, side by side with the antinuclear movement.

Around twenty million people in the United States joined demonstrations and community events to celebrate the very first Earth Day on April 22, 1970. The Environmental Protection Agency (EPA) was founded at the end of the same year. Citizens were pissed off about the state of the planet, and rightfully so. For those of us lucky enough to grow up in a post-EPA era, it's hard to understand just how polluted the United States had become. Images of smog-filled cities and flaming rivers had Americans boogying on down to eco-centric protests and marches. All that noise made a difference, and by the early seventies the Clean Air and Clean Water Acts were . . . cleaning up our act. The EPA banned DDT and brought bald eagles back from the brink, got rid of leaded gasoline, and started protecting species teetering on extinction.

Emboldened by these wins, well-meaning environmentalists shifted their focus to another problem: atomic weapons testing.

On a sunny but chilly day in the fall of 1971, a group of activists and journalists boarded a small fishing boat off the coast of Vancouver in Canada. The vessel, typically used for catching halibut in the cold waters of Alaska, was about to enter history by playing a key role in the birth of one of the biggest movements ever. As the group sailed off to Amchitka Island, part of the Aleutian Islands in Alaska, they hoped to make the two-thousand-mile journey on rough waters and get there in time to stop a devastating act from taking place.

Remember when we talked about the Partial Test Ban Treaty that prohibited aboveground testing? Well, here's the thing . . . it still allowed *underground* testing.* Baby steps, I guess. By 1970, the United States military had conducted two underground tests on Amchitka Island, which people feared had the potential to trigger earthquakes and tsunamis. When the government announced a third test to occur in the fall of 1971, a group of seven thousand activists decided they'd had enough and organized against it. They created the Don't Make a Wave Committee and came up with a plan to charter a boat and sail into the test zone to confront the bomb. Bold. Chaotic. Peak protest energy. They named the boat *Greenpeace*.

The passionate activists weren't able to stop the test. But they were able to raise awareness and inspire others to join in protest against nuclear weapons testing. Not bad for a bunch of folks on a fishing boat. The following year, the Don't Make a Wave Com-

---

* The treaty allowed underground nuclear tests as long as the radioactive debris remained within the country's borders. It prohibited nuclear weapons tests in the atmosphere, in outer space, and under water.

mittee officially changed its name to Greenpeace, a nod to the chartered fishing boat used in its first demonstration. Greenpeace started out as an anti–*nuclear weapons* group, but its mission quickly expanded to broader issues like whaling, deforestation, and pollution. Unfortunately, its opposition later expanded to include nuclear *electricity*. It remains one of the most successful and well-funded environmental organizations in the world.

The environmental movement was incredibly powerful in rallying people around a good cause. But like any large-scale effort, it created strong opinions and didn't leave much space for nuance. On one side were the crunchy, happy, natural, planet-loving protesters, and on the other, the energy-guzzling, corner-cutting industrialists. In an age when fossil fuels were by far the main source of energy, drawing the lines that way was fair. Because of its history, nuclear electricity ended up on the wrong side of this divide, which means that many of our oldest and most trusted environmental groups are against it.

Greenpeace and other organizations overstated the dangers of radiation to get people to push back on proliferation. It's true that we knew less about radiation at the time. But we're living with the fallout of that fear today—despite the fact that nuclear power plants have proven time and again that they're extremely safe.

The Natural Resources Defense Council (NRDC) was another influential entity that protested nuclear power plants starting in the 1970s. The NRDC was founded in 1970 and quickly became involved in efforts to address all sorts of environmental issues. Its opposition to nuclear electricity came from concerns about safety, nuclear waste disposal, and the potential for accidents like the Three Mile Island incident in 1979.

We'll take a quick pit stop here to talk about the Three Mile Island accident. If you've heard of it before, you might think it proves that nuclear electricity is dangerous and bad. But the details aren't as horrifying as most people have been told. Yes, a combination of glitches, human errors, and bad luck led one of Pennsylvania's Three Mile Island reactors to overheat and partially melt. Not ideal. Despite the drama, the plant's containment structure successfully prevented any major radiation leaks.

But, there's a "but." Some radioactive gases *did* escape, and people in the area got spooked. That part is understandable; no one *wants* to hear that their local nuclear reactor got a little too steamy to function.

The thing is, no one got hurt. Zero injuries or deaths. In the grand scheme of nuclear electricity history, this was a hiccup, not a catastrophe. It was a wake-up call for making plants even safer, and it also showed that when things do go wrong, a reactor can avoid a disaster.

Of course, the media fanned and spread the flames of that fear, creating a PR nightmare for nuclear electricity. It didn't help that the accident happened less than two weeks after the premiere of *The China Syndrome*, a disaster thriller starring Jane Fonda as a television reporter who reveals that a local nuclear power plant is covering up numerous safety issues. The title of the movie comes from the idea that a nuclear meltdown would cause components of a reactor to sink into the ground and go all the way through the planet, finally reaching China (lol). *The New York Times* reported that nuclear experts called the film fiction and a "character assassination of an entire industry" because, well, it was.[3] Just imagine how damning it looked when a good ol' nuclear plant had an accident *suspiciously similar* to the one in the movie just days later. You could practically hear people shouting "I told you so."

Inspired by the Three Mile Island incident, well-known musicians including Jackson Browne, Graham Nash, Bonnie Raitt, and others formed Musicians United for Safe Energy (MUSE). They organized a weeklong series of concerts called "No Nukes" at Madison Square Garden in New York City. It featured high-profile artists who supported the cause, like Bruce Springsteen and James Taylor. Almost two hundred thousand people attended the No Nukes rally, making it one of the largest antinuclear demonstrations in U.S. history.

It's hard to overstate the cultural impact of getting all those artists, some of whom were at the absolute top of their game, together in one place to decry the dangers of nuclear everything, even electricity. My favorite part is when one of the bands on stage sang about wanting the "comforting glow of a woodfire" but not "atomic poison power." Which is ironic, considering that burning wood emits toxic pollution and greenhouse gases. While only 4 percent of residential heating in the United States comes from wood-burning stoves, their smoke contributes to about 6 percent of all fine particulate matter emissions.[4] Some studies estimate that wood-burning stoves could cause up to forty thousand premature deaths a year in the United States alone, and they likely cause millions of deaths worldwide.[5] But sure, keep on chasing that "comforting glow."

When the Chernobyl disaster followed in 1986, the negative image of nuclear electricity became firmly entrenched in the public's mind. This image wasn't based on science, but it stuck. In the decades that followed, it kept brilliant people from choosing a career in the nuclear industry, something that's biting us in the ass today. A 2019 report revealed that nuclear electricity was still struggling to get rid of outdated perceptions that it's a dying industry and faced a serious shortage of up-and-coming skilled laborers.[6] As of 2024, there are more jobs than workers in nuclear electricity, and the folks trained to do those jobs are rapidly aging out. Various schools and organizations are working to pump some life back into training and recruitment, but there's no doubt that one of the hurdles is that students grew up hearing that nuclear electricity was bad, outdated, and on its way out. Not exactly a hot career pitch.

Politicians historically got points with constituents for closing nuclear plants, even though it always led to higher emissions. Most places that shut down nuclear plants have to replace their electricity output with coal or gas. We saw this happen in real time in New York, Massachusetts, and Pennsylvania.[7] After closing their nuclear plants, they each saw a rise in greenhouse gas emissions. New York's CO_2 emissions climbed almost 15 percent after shutting down just one nuclear power plant, Indian Point.

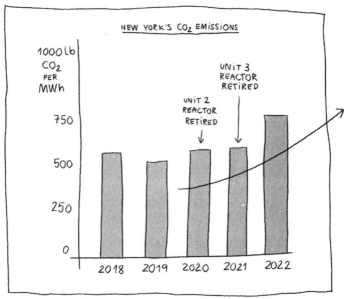

SOURCE: THE GUARDIAN, 2024

Antinuclear sentiment even snuck into people's homes without them knowing. Take *The Simpsons*, for example. If you're too young to know that Homer's soul-sucking job is in the nuclear industry, your parents do. This single show—which has earned about a million views per episode since it aired in 1989—played a significant role in convincing multiple generations that nuclear electricity is a sloppy, dangerous industry that creates toxic spills and causes cartoony mutations.

Back in 2009, William Irwin, a philosophy professor at King's College and the editor of books about television and pop culture such as *The Simpsons and Philosophy*, told reporters that he felt the show had increased many viewers' concerns about nuclear electricity. He pointed out that Homer's bumbling personality didn't inspire great confidence in reactor operators, recalling an episode where the good-hearted buffoon had pressed random buttons to stop a meltdown.

Irwin also noted that Lisa, who's obviously the show's smartest character, is staunchly antinuclear. "She's very eco-friendly, and very much against nuclear power and the nuclear power plant run by Mr. Burns," he told a Canadian radio station in 2009.[8] Personally, I feel like the most damaging aspect of the show's spin on nuclear (other than Mr. Burns being evil) is all the *goo* that's supposed to represent nuclear waste. Whether it's packed away in rusty metal barrels or actively leaking out onto the floor of a storage facility, the image of nuclear waste as an uncontainable, sticky-icky slime has done heaps of damage to nuclear's public image. We've already gone over what spent fuel looks like, but here is another image so you don't forget:

Paris Ortiz-Wines and Mark Nelson in front of nuclear waste casks. Courtesy of Paris Ortiz-Wines.

We can't pin all our failed nuclear potential on Homer Simpson, though. The creative geniuses behind *The Simpsons* got their antinuclear bias from somewhere else. If only we could figure out who was behind such a dastardly deed. Oh, wait. We can.

GRUBBY HANDS

To no one's surprise, the fossil fuel industry was all about the antinuclear movement. Remember how oil money helped kick-start the antinuclear group Friends of the Earth? The fossil-fuel-to-antinuclear pipeline never stopped.

In the 1980s, Exxon, a big player in the oil and gas game, poured funds into groups opposing nuclear electricity like Friends of the Earth and the NRDC. Its aim was to create doubt about climate science and, surprise, surprise, to promote the idea that nuclear wasn't a viable solution.

In 2012, it came out that the Sierra Club, a grassroots environmental organization, had accepted $25 million in donations from the methane gas industry between 2007 and 2010.[9] In 2007, the Sierra Club's director of global warming was quoted saying, "Switching from dirty coal plants to dangerous nuclear power is like giving up smoking cigarettes and taking up crack."[10] Total coincidence, I'm sure.

And check out this ad in a Long Island newspaper from the 1970s. All quaint and unassuming at first glance. "LILCO [a utility company] is building a nuclear plant in your backyard." It features a creepy-looking sun next to the words "Solar, Not Nuclear." Aw, wholesome! Until you squint at the fine print and see it was *paid for by the Oil Heat Institute*. Why would the oil industry be advocating for solar power instead? Could it be because they know that nuclear electricity would put them out of business?

But my all-time favorite is an ad from Australia that ran in 2007. It has a photo of coal miners looking rather sad and urges workers to support a bill that would invest AUD $1.5 billion in "clean coal technology," which is an oxymoron. And right there, in bold text, they just say it: "Nuclear Power Will Kill the Coal Industry." I mean, thank you for the honesty? The coal industry *knows* that nuclear electricity would put them out of business and they're determined to keep it on the sidelines.

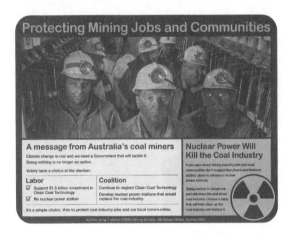

In recent decades, the fossil fuel industry realized they can stay in the game a bit longer if we move toward intermittent sources of energy that need backup. Methane gas companies began selling their product as the perfect partner for renewable energy—and to be fair . . . it kind of is. Methane gas peaker plants,*

---

* Gas peaker plants are designed to run only when there's more demand for electricity than other sources of energy, typically renewables, can provide. Let's say a town has a bunch of solar panels and a gas peaker plant. On a cloudy day or after sunset, the peaker plant would kick in and make sure people have energy when they need it, not just when the sun is shining.

designed to run only when needed, can ramp up or down quickly and provide backup when renewables inevitably go down. This crafty collab keeps us tethered to fossil fuels, allowing companies to greenwash their products as essential for the clean energy transition.

As troubling as that is, the story gets even darker. Grab the popcorn and pop a Xanax, because what comes next is a master class in corporate manipulation that just might raise your blood pressure. Back in 1999, some Greenpeace activists decided to set up their own independent energy provider—a cooperative in Germany that promised customers sustainable energy from solar and wind. It ran under the Greenpeace brand for years, but in 2021 it was quietly rebranded as Green Planet Energy.

In 2011, this cooperative rolled out a product called "proWindgas." You might be confused because pro-wind . . . gas? is very much not a thing, so what could this magic product be? It turned out to be nothing more than imported methane gas—the fossil fuel variety. When criticized, the cooperative defended itself by saying the plan was to *eventually* mix the methane with hydrogen produced from renewable energy. By 2021, it had managed to ramp up the hydrogen share to an impressive 1 percent, with the rest of the mix consisting of 10 percent biogas (burned trash and literal shit) and 89 percent methane.

So, yeah, the reason Greenpeace's energy cooperative rebranded to Green Planet Energy is because it got busted selling Russian methane gas under the greenwashed name "proWindgas."

In 2015, the energy cooperative tried to sue the European Commission over approving state aid to Hinkley Point C, a nuclear power plant in England, citing it as a potential competitor on the energy market.[11] Thankfully, sanity prevailed, and the European Court of Justice denied the cooperative's request as inadmissible.

That kind of shady nonsense is still going on everywhere. In 2019, *The Daily Beast* showed that the American Petroleum Institute (API) helped fight nuclear electricity investment in Pennsylvania.[12] Proposed House Bill 11 named nuclear as one of the energy sources that utility companies could count as clean, but API paid for social media ads and mailers warning voters to vote against it, so it never passed. They plotted to use environmental activism and social justice as a smoke screen to keep fossil fuels burning. Just imagine all the stuff we *don't* know about.

Consider this the plot twist moment in the book.

I'm about to say something that's going to sound unhinged, especially after everything you just read:

I'm glad the antinuclear movement slowed down the deployment of nuclear electricity.

Yes, even though it prolonged our fossil fuel addiction.

Yes, even though it made climate progress harder.

And no, I haven't lost the plot. Hear me out.

Unlocking nuclear fission was humanity's entrance into its main-character era, the point of no return. We transcended our humanity and gained powers previously reserved for the gods: the ability to create and destroy at a scale never seen before. Naturally, this freaked people out. Splitting atoms? How is that even real? How did we manage to harness energy from something we can't even see? It all sounds completely insane, feels like we accidentally cheated our way into forbidden god-mode. We bit the apple, and the apple can now kill us all. This had profound psychological impacts on civilization and triggered deep existential angst, creating a spiritual crisis of sorts. We had to come to terms with this newfound power and wonder if we were wise enough to have it. Hint: We were not.

Humanity had to grow up fast and grapple with the complexities of wielding nuclear technology. This potential, and the responsibility that came along with it, needed to be reckoned with. The antinuclear movement *forced* us to slow down, giving us much-needed time to get it right. As a result, we stopped atmospheric weapons testing for decades, pushed for disarmament, and made nuclear electricity one of the best and safest energy sources on Earth.

Now that I am done being nice, let's get something straight: While opposing nuclear electricity kind of made sense once upon a time, doing so in 2025 is outdated and flat-out dangerous.

TL;DR

Our reckoning with nuclear electricity has been a wild roller coaster. From absolute delusion to terror, we've felt every emotion in between. Shout-out to those early environmentalists for kick-starting the environmental movement in the 1970s and cleaning up our air and water. Sadly, they lumped nuclear electricity in with fossil fuels as the "bad guys." Even Hollywood jumped on the hippie bandwagon and painted the technology as evil and dangerous at every opportunity. No, nuclear reactor operators aren't dumb like Homer Simpson. No, nuclear waste is not green goo oozing out of barrels. And no, the components of a nuclear reactor can't travel through the center of the earth all the way to China in case of an accident. You know who loved this assassination campaign? The fossil fuel industry. They've tried to get nuclear out every step of the way and position their toxic products as the perfect partner for renewable energy.

8: THREE DECADES OF BAD ENERGY POLICY

A TALE OF TWO DECARBONIZATIONS*

If you want to know what *not* to do to create a rad future, just look at Germany—the land of bratwurst, solar power, and questionable decisions. It seems like Germany is hell-bent on being *consistently* on the wrong side of history. In April 2023, it shut down its last nuclear reactors, after a multidecade antinuclear crusade called the *Energiewende*.† They put solar panels on every roof and wind turbines in every field, shouting *"Energiewende!"* from the mountaintops. Cute, no doubt, but let's just say their energy bills could make anyone say *"scheiße."* The *Energiewende* is projected to cost over €1.2 trillion by 2035.[1] Despite pouring a bunch of time and cash into

---

* "Decarbonization" is the process of reducing or eliminating carbon dioxide (CO_2) emissions.

† The *Energiewende* is Germany's epic (and pricey) quest to ditch nuclear and fossil fuels for wind and solar, bravely facing blackouts, high bills, and a love-hate relationship with Russian gas.

Three Decades of Bad Energy Policy

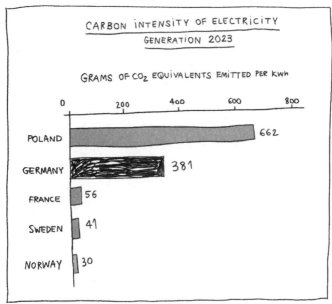

SOURCE: OURWORLDINDATA.ORG

this vision, Germany still has one of the dirtiest electrical grids in Europe.[2]

But there's a better way. Let's learn how with the tale of two decarbonizations.

On one side of this tale, we've got Germany. Back in the 1970s, the government was super into nuclear electricity, while citizens had some reservations. When the quaint hamlet of Wyhl was chosen to host a nuclear reactor in 1975, local winemakers flipped out. The opposition grew into something bigger when their protests were met with police violence, which got media attention and drew supporters from all over Germany. By 1981, plans to build new nuclear power plants were met with protests one hundred thousand people strong.

Then came the accidents at Three Mile Island and Chernobyl,

which dampened federal enthusiasm. The Chernobyl disaster hit the Germans especially hard, as people worried about radioactive contamination in their food and environment. "That's a memory that stays with people, if you remember that you can't let your children play outside because you're scared of radioactivity," said Miranda Schreurs, a professor at the Technical University of Munich, in a 2022 *Politico* interview.[3] As we've talked about, these fears were overblown and not based on facts. The Germans expected a spike in cancer cases that never materialized.

Germany's trepidation about nuclear electricity, however understandable, had broad consequences. It stopped building new plants in 1989, and by 1998, phasing out nuclear had become a core piece of national policy. As far as Germany was concerned, going green was as much about kicking nuclear to the curb as it was about fighting pollution and climate change. Its environmental movement mirrored the one in the United States, with the German Green Party, founded in 1980, rising out of the antinuclear, environmental, peace, and social movements of the late twentieth century.

In a moment of sanity, Angela Merkel's government in 2010 decided to keep some of the country's remaining nuclear plants open—naturally stirring up antinuclear activists. Protests erupted, from tens of thousands surrounding Merkel's office in September to massive demonstrations in Munich a month later and violent protests against a train carrying reprocessed nuclear waste in November. Then came the Fukushima disaster in March 2011, and Merkel's nuclear policy did a one-eighty.

A few months later, Merkel admitted, "Before Fukushima . . . I was convinced that it was highly unlikely that [an accident] would occur in a high-tech country with high safety standards . . . Now it has happened." French president Nicolas Sarkozy was stunned. "I tell her—but Angela, what's going on? How can this be?" he said in a parliamentary hearing, recalling their conversation. "She says, but Nicolas, have you not seen Fukushima? And I said—but where is the tsunami going to come from in Bavaria?"[4]

It didn't matter that tsunamis are nonexistent in landlocked Bavaria, or that one accident doesn't justify discarding an entire technology. Politicians want to be popular. Had it not been for massive protests the year before Fukushima, Merkel's response might have been less drastic.

But history took its course, and the German government committed to fully phasing out nuclear electricity by 2022.

To drive home the absurdity here: Germany postponed the shutdown of its last three reactors from December 2022 to April 2023—not because the government realized scrapping nuclear was a terrible idea, but because the Ukraine war slashed its Russian gas supply.[5] Without Russian gas *and* nuclear, the German grid wouldn't be able to stay afloat. The most nefarious act in this circus is that Germany's former chancellor, Gerhard Schröder, sat on the boards of not one, not two, but *three* Russian energy companies. By 2022, he was pocketing a chill $1 million a year for doing so.[6] By the start of the Ukraine conflict, Germany was paying Russia about $220 million *a day* for gas to keep its "nuclear-free" dream alive.[7] They knew they were replacing perfectly good nuclear power plants with fossil fuels.

Germany is touted as a leader in solar, and its early investments helped make it one of the cheapest sources of energy.[8] There's just one problem. If you have been to Germany, you know it's not . . . how can I say this? A very sunny place. In sunny spots like California, solar has a 24 percent capacity factor. In Germany, it's a depressing 10 percent—meaning solar panels are only cranking out electricity at full capacity 10 percent of the year. Not to be rude, but that is extremely lame. Germany wanting to be a leader in solar is like Sweden trying to be a leader in pineapples.

Ironically, German reactor designs and engineers were among the best in the world. Had they maintained focus, they could have dominated the market for affordable nuclear tech. Instead, Germany chose to rely on coal and imported gas, making its electricity nearly ten times dirtier than France's.

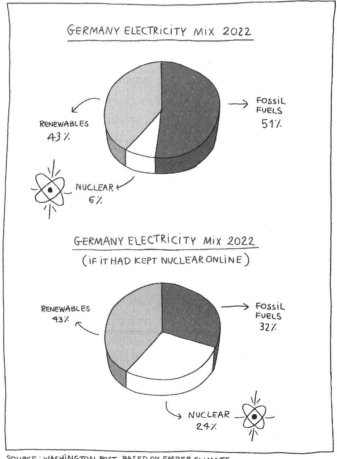

SOURCE: WASHINGTON POST, BASED ON EMBER CLIMATE

What would have happened had they kept their nuclear plants online? We don't have to guess. The illustration on the previous page shows what Germany's electricity could have looked like.

Fatih Birol, the head of the International Energy Agency, has called Germany's nuclear exit a "historic mistake."[9] Because it is.

To make things worse, Germany isn't just delaying its own transition to clean energy—it's trying to convince the rest of Europe to join the slow train. In 2023, Germany and six other countries argued that hydrogen made with nuclear should not count toward EU renewable energy targets.[10] Meanwhile, burning trees and poop does.

(I reserve the right to say nice things about Germany in the next edition of this book. Because as it's heading to print, I am hearing rumors that Germany might finally do the right thing and turn nuclear reactors back on.)

Sorry about the bad vibes. Let's hop over to the happier side of our decarbonization tale for a bit of a palate cleanser. We're going to France!

THE LAND OF WINE, GREAT AESTHETICS, AND NUCLEAR ELECTRICITY

In the 1970s, France was hooked on foreign oil—about 75 percent of its energy came from it. When the 1973 oil crisis hit* and petroleum nations decided to choke the global supply, France found itself scrambling to keep the lights on. That's when French prime minister Pierre Messmer stepped up and launched the audacious

---

* Just to remind you, this is when the biggest oil countries slapped an embargo on oil exports to the United States and its allies, including France.

Messmer Plan. This wasn't just a plan, it was a nuclear revolution. Messmer's vision was to free France from the grip of foreign oil by making the country nearly energy independent, powered by nuclear ingenuity. And it was a success: In just fifteen years, France built forty-eight reactors. In 2024, it had fifty-six reactors producing about 65 percent of the nation's electricity. France went from fossil fuel junkie to nuclear powerhouse in a single generation. Thanks to its nuclear-heavy energy mix, France has one of the cleanest grids among developed countries, with low carbon emissions and a stable, reliable supply. French electricity is also cheaper than its neighbors', which allowed the country to be the largest net exporter of electricity in Europe in 2023.[11]

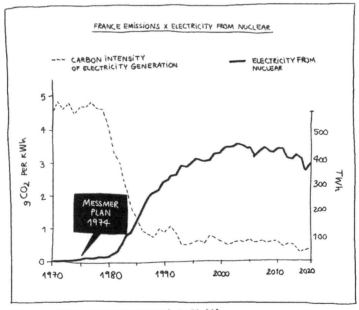

SOURCE: NATURE ENERGY & OURWORLDINDATA.ORG

Despite proving that nuclear electricity can be scaled fast and replace fossil fuels, even the land of baguettes fell briefly for antinuclear propaganda. In order to hit an arbitrary "renewable energy" target, in 2012 soon-to-be-president François Hollande promised to cut France's nuclear share from 70 percent to 50 percent of all electricity. The goal wasn't to reduce the country's carbon emissions; the goal was to increase its share of renewables.[12] That's silly because they would be replacing reliable, clean electricity from reactors with *less* reliable, clean electricity from solar panels or wind turbines, which would ironically lead to higher emissions.

In 2024, Emmanuel Macron's government changed course and France is back in the nuclear electricity game. It plans on building six more reactors, with an additional eight under consideration. *Oui!*

WHO KILLED NUCLEAR ELECTRICITY?

Pop culture and the persistence of the antinuclear environmentalist movement no doubt had an impact on people's feelings about nuclear electricity. Yes, Jane Fonda *is* an icon, and *The Simpsons* had a wide reach. But it would be silly to think a bunch of hippies, a few celebrities, and some hit cartoons were all it took to bring down a once-thriving industry.

What really happened? Why have we not been building new nuclear reactors left and right in the United States?* Well, as with anything, the answer is complicated. Let's investigate.

---

* I will be focusing on the United States for this part. Partly so as not to bore you, partly because it serves as a good example for other places.

In 1946 the United States created the Atomic Energy Commission, a government agency that oversaw nuclear electricity. The agency supported a wave of new nuclear reactors. Said reactors were all expensive at first, as is the case with any new tech. But in the following fifteen years, companies figured out how to make them cheaper (spoiler: by making them bigger).

All-American nuclear was crushing it. Nuclear reactors offered some of the cheapest electricity around, at one point becoming even cheaper than modern natural gas plants. Unsurprisingly, utilities started ordering more and more of them. But that's when things started going south. The supply chain for all the components that go into a nuclear power plant wasn't very robust or mature and became stressed, causing delays. Delays made it easy to paint nuclear reactor projects as being unpredictable in terms of cost and timing. Regulations for nuclear plants were also changing quickly during this time, so the government could upend construction or operations on a whim.

Imagine you're building a house with a set budget. You hire an architect, a general contractor, and a full construction crew, and they start placing orders for essentials like pipes and concrete. But these materials keep getting delayed due to random issues. Meanwhile, you're still paying the entire crew—even when they're just sitting around waiting for deliveries.

Then the city steps in with new building codes. Just as your long-awaited pipes finally arrive and get installed, you find out they're now banned under these new rules. So, you pay your crew to tear them out, order the new (now-approved) pipes—likely to be delayed too—and reinstall them. By now, your dream home is costing a lot more than you ever planned.

With shifting regulations and supply chain setbacks, reactor costs had already more than doubled by the early 1970s. Then came the 1973 oil crisis. While it caused an uptick in demand for nuclear electricity, it also drove up inflation, which spiked nuclear plants' costs because of how they were financed. Utilities borrowed money to build plants, so rising interest rates sent costs soaring. When the oil crisis ended and fossil fuels became cheap again, nuclear suddenly seemed unnecessary. The final blow came in 1979 with the Three Mile Island accident. Although no one was harmed, it ignited massive antinuclear protests, and utilities got spooked. The fifty-one reactors under construction faced even more delays as regulators changed safety protocols, sending costs through the roof. Then things got really bleak: All 120 reactor orders were canceled, one after another, and not a single new reactor was built from 1978 until 2013.

And you know who benefited from this? The fossil fuel industry, obvi. The United States went from generating 72 percent of its electricity from gas, coal, and oil in 1985 to 87 percent by 2013. That's thirty-five years when the entire nuclear electricity industry failed to grow at all—in fact, plant shutdowns made it *shrink*. Just imagine the impact of this standstill on supply chains, education, and legislation—hurdles that today's nuclear industry is still trying to overcome.

The good news: In 2024, the United States finally completed its first reactor in thirty years (yay . . . ?) at the Vogtle Electric Generating Plant in Georgia. The bad news: The process didn't exactly get utility companies hyped about nuclear electricity. The reactors took more than seven years longer to build than planned and cost $36 billion—more than twice the original budget.

Besides inevitable COVID-related delays from supply chain and worker shortages, Vogtle was plagued by a lack of experts, frequent design changes, and some serious project mismanagement. Believe it or not, they started construction before the design was fully complete.[13]

This brings us to a tough truth for nuclear supporters: Yes, public opposition to nuclear was based on plenty of misinformation and had negative consequences. But the Western nuclear industry also hasn't delivered when it comes to finishing projects on time and within budget. Sophisticated critics know this, so instead of asking "What about Chernobyl?," they point to these modern-day cost overruns to oppose nuclear for being too expensive.

Other countries have done a better job, like China, Russia, South Korea, the United Arab Emirates, and Japan. In 2024 the UAE successfully finished building four reactors at the Barakah nuclear power plant. It took them twelve years, but now Barakah alone makes 25 percent of the country's electricity. South Korea not only kept costs under control, it also made building reactors cheaper as time went on. The Koreans started their program a little later than other countries, so they were able to use designs that had already been tested and modified. Korea also created a stable regulatory body—one that wasn't like a boyfriend who sends mixed signals and makes you question your sanity. It picked a single reactor design and stuck to it, which allowed workers to get better at building with each new project. That's easier for countries like South Korea, because it only has one utility company, which is owned by the government. Additionally, South Korea also stuck with the same construction company for its projects. Compare that to the United States, where there are thirty power companies, who also act as project managers and choose different designs.

By the way, Americans have successfully built cost-effective reactors in the past. St. Lucie Unit 2 in Florida was a successful case. This reactor was started in 1977, seven years after the twin plant's first reactor, and took about five years to complete. It cost $1.45 billion ($4.4 billion in 2023 dollars), financing and all, despite having more steel and concrete than a modern reactor design. This smooth construction process wasn't because people were in love with nuclear electricity at the time either, as the Three Mile Island accident happened less than halfway through construction. The truth is that St. Lucie managed to avoid all the wishy-washy project management issues that happened later. Florida Power and Light, the utility company that built it, had just completed the construction of another three reactors, so its staff had become experts on the process. They proved that having a good project management team could alleviate all sorts of potential problems. That's not unique to nuclear electricity—it's just common sense.

At its core, nuclear reactors aren't inherently expensive. It's not like they're made of unicorn eyelashes, they're just a big pile of concrete and steel with some clever engineering. Some of the high costs come from a mix of valid concerns and exaggerated fears that have made nuclear one of the most heavily regulated industries out there.

But while those regulations slow things down and add expense, they don't fully explain why the industry has struggled so hard. Another reason why operating nuclear power plants is more expensive than, say, a solar farm, is because they require lots of workers who earn good salaries. Call me crazy, but I think this is a good thing. Regardless, it's time for the Western nuclear industry to step up and get back to its prime.

TL;DR

Germany sucks. Just kidding. But seriously, Germany had the potential to be an energy-independent technological leader thanks to its early investments in nuclear electricity. Instead, it shut down perfectly good nuclear power plants in a bid to appease the misguided antinuclear crowd. It burned a lot of fossil fuels (and cash) as a result and yet still has one of the dirtiest electrical grids in Europe. France, on the other hand, showed with elegance what it looks like to grow one's economy while reducing emissions from electricity production. Be like France. After decades of stagnation, nuclear electricity has become one of the most expensive sources of energy in America. But it doesn't have to be. We can learn from countries who have managed to build reactors quickly and cheaply: Stick with a single proven design and deploy lots of it.

9: VIBE SHIFT

My heart races as I quickly open the door of the black rental car and sit nervously on the passenger side. As I lower the sun visor and open the mirror, my wide eyes stare back at me with excitement. With shaky hands, I place blue eyeliner stickers on top of my eyelashes and take a couple of deep breaths. Black bullhorn in hand, I step out of the car and into one of the most memorable moments of my life. It was time to launch ISODOPE into the real world.

A year earlier, I realized that just posting facts online wasn't going to cut it. If I wanted to start a real revolution, I'd have to move beyond changing minds—I needed to change atoms. It was 2021, and five U.S. nuclear plants were on the chopping block to be closed before their time, mostly because of politics. Saving them was the low-hanging fruit to keep emissions from spiking and slam the brakes on climate change. History shows when nuclear plants shut down, fossil fuels usually step in to fill the gap. Of the five, Diablo Canyon stood out, perched on a bluff above the Pacific Ocean near gorgeous Avila Beach, California.

Diablo Canyon, California's last standing nuclear plant, has the wildest lore marked by unlikely alliances and a legacy that made it the ultimate symbol of America's antinuclear movement. Back in the 1960s, Pacific Gas & Electric (PG&E), the big-shot utility that owns it, cut a weird deal with the Sierra Club to build the plant on a pristine coastal strip just twelve miles west of San Luis Obispo. PG&E was originally set to build five reactors down at Nipomo Dunes, a sandy coastal gem south of town. This plan was met with fierce opposition from the Sierra Club, who wanted to protect those dunes at all costs. In exchange for the club's cooperation, PG&E gave up on its plans for the controversial Nipomo Dunes site and chose Diablo Canyon instead. Believe it or not, the Sierra Club organization wasn't against nuclear electricity back then. Its president, William Siri—a physicist with a background in nuclear medicine—believed nuclear electricity could be an environmentally friendly alternative to fossil fuels.

But the deal triggered drama within the Sierra Club and led to a messy breakup with its first executive director, David Brower. As we dug into in chapter 6, Brower was a celebrated mountaineer and conservationist who had been a big deal in the club, growing their membership from a modest seven thousand to a whopping seventy thousand between 1952 and 1969. Furious over the club's support for Diablo Canyon, he left to start his own thing: Friends of the Earth.

Interestingly, Brower's issue with nuclear wasn't even about safety. His main beef was that cheap, abundant electricity would fuel a population boom in California, spoiling its natural beauty. Oh, and Brower was hardcore anti-immigration too, adding another layer to his uniquely awful worldview.

While Diablo Canyon was being built, nearly two thousand activists, including famous musician Jackson Browne, were arrested for chaining themselves to the gates and stopping workers from getting in. That's by far the largest number of arrests in the history of the American antinuclear movement.

After years of hiccups, screwups, and nonstop protests, the plant's two reactors finally kicked on in 1985, pumping out clean electricity despite everyone and their mom trying to kill it. As a part of the anti-Diablo campaign, a smaller No Nukes concert took place in San Luis Obispo in 1979, a little brother to the Madison Square Garden blowout. Same lineup of musicians. Same misinformation. About thirty thousand people showed up to yell antinuclear chants. The highlight came when Governor Jerry Brown hopped onstage and promised he would fight to shut Diablo Canyon down, leading the crowd in a chant of "No on Diablo! No on Diablo!"

Brown didn't stop there. In the following three years, he and his allies killed numerous nuclear projects across California. If those plants had been built, California's electricity would be nearly all clean today. To make things spicier, it turns out Jerry Brown's dad made a good chunk of his fortune by . . . lobbying for and selling Indonesian oil and gas in the United States.[1] Pure coincidence, I'm sure, nothing to see here.

Fast-forward to 2016, and Brown, in his fourth term as governor, was finally on track to fulfill his "No on Diablo" promise. PG&E, NRDC, Friends of the Earth, and other organizations hashed out a deal to close the two reactors by August 2025. Brown swore, like every politician with a microphone does, that Diablo's output would be replaced by renewables.

Before the deal was signed into law, Michael Shellenberger, a journalist and environmentalist who ran an organization called

Environmental Progress, led an effort to keep the plant open. He connected with Heather Hoff, an engineer at Diablo Canyon and passionate mother of a young girl, who wanted to get involved. Shellenberger gave a no-bullshit talk to plant employees, in which he clarified that without Diablo Canyon, California wouldn't meet its climate goals, adding that the plant could stay open for at least another twenty years. He knew Diablo Canyon was being closed for political nonsense and encouraged the workers to organize to save their plant.[2]

Inspired by Shellenberger's talk, Hoff connected with another plant employee, Kristin Zaitz, also an engineer and mother of young children. They started their own organization called Mothers for Nuclear, a global community of mothers in support of nuclear electricity.

In the summer of 2016, Shellenberger, Hoff, and Zaitz led the March for Environmental Hope, hauling advocates from San Francisco to Sacramento's state capitol. It was the first pronuclear march in U.S. history and featured rallies outside the offices of the Sierra Club, Greenpeace, and the NRDC, all organizations behind Diablo Canyon's closure. Their efforts didn't pan out. Governor Brown slapped his signature on the bill in September 2018.

This was a huge blow to the pronuclear movement, causing them to lose faith that the world would ever embrace nuclear electricity. By the time I created a *new* campaign to save Diablo Canyon in 2021, it was almost impossible to get folks involved. I talked to as many pronuclear organizations as possible, but nobody seemed to think it was worth the fight. Everyone was jaded. "It's a done deal," one told me. "Lots of people have died on that hill; don't do it," said another. But I had an intuition that this time it could be different, because California's energy landscape had completely changed since 2016.

Diablo Canyon alone had generated around 9 percent of California's electricity in 2020, making it the state's *largest* source of carbon-free energy. By then, California had set a goal of 100 percent clean electricity by 2045. You don't need to be a math genius to see that knocking out its biggest source of clean energy wouldn't be helpful. Shutting down Diablo wouldn't just derail those climate goals, it would also increase the risk of blackouts—which turns out are not very popular with voters.

In 2020, Governor Gavin Newsom issued an executive order mandating that by 2035, all new cars and light trucks sold in the state be electric. More electric vehicles would mean a massive jump in electricity consumption, especially at night, since that's when most people plug their cars in. You know what doesn't produce power at night? Solar panels. You know what does? Nuclear plants.

At the time, California was already struggling to keep the lights on. To prevent blackouts during scorching summer days, the state hit residents with "flex alerts" begging them to dial down air-conditioning and other electricity-guzzling activities. They even cut power in certain areas to balance out supply and demand and prevent the grid from collapsing.

California's grid struggles, combined with its aggressive clean electricity goals, made me realize there was still a chance to save Diablo. Eventually, I found my people. A ragtag crew of nuclear misfits and clean energy rebels: Paris Ortiz-Wines from Stand Up for Nuclear, Mark Nelson from Radiant Energy Group, and the badass duo Hoff and Zaitz from Mothers for Nuclear. We were loud, relentless, and borderline obsessive, but we were scattered. What we needed was direction and a plan that could turn our passion into something that might actually move the needle.

After months of Zoom calls and brainstorming, I realized the key was to convince Governor Newsom that keeping the plant open was a no-brainer. Bigger than that, we needed to show that the public would be cool with the decision. In one of those moments that make you believe in destiny, a couple of days later, I heard through the grapevine that we weren't the only crazies after all. Clean Air Task Force, a legit climate and environmental organization, had encouraged brainiacs at MIT and Stanford to work on a pivotal study. They were analyzing the possibility of keeping Diablo Canyon open and the impact that would have on California's energy costs, reliability, climate goals, and land.

Later, Clean Air Task Force helped form and advise Carbon Free California, which mounted polling, public affairs, and a proper lobbying campaign, encouraging state leaders to support the continuation of the plant.

Inspired by the renewed momentum, I decided to shake things up and throw a rally in San Luis Obispo. But there was a problem. I had no idea how to pull that off. Where do you go? How do you get people to show up? What are you even supposed to do at a rally? Do I need to bribe people with free pizza? (The answer is always yes.) I had never organized *anything* like that in my life.

In moments like these, when I was biting off more than I could chew, I thought back to my grandmother Dida. One of twelve children, she was born in the 1930s in a rural area on the outskirts of a small town in southern Brazil. She had a passion for learning, especially science. Her parents ran a roadside market and a small restaurant, where locals on horseback or passing drivers would stop to stretch their legs and get a bite to eat. Somehow, she figured out how to buy books through the mail and taught herself physics and math, spending hours alone reading underneath a fig tree. She eventually became a dentist, an unheard-of profession for a woman in that period and that part of the country. She also raised five children. Her determination and willpower have been an inspiration throughout my life. If my grandmother could do all that in the freaking 1930s without any tech, I could surely figure out how to organize a rally.

Somewhere along the process, I befriended Carolyn Porco—yes, *that* Carolyn Porco. The planetary scientist whose tweet about molten salt thorium nuclear reactors had lodged in my brain years earlier. She happened to live in California. A highly respected scientist, Porco had also been a strong nuclear electricity advocate for over a decade. What's more, she had to fight antinuclear hysteria throughout her career. In a March 2024 interview, Porco explained how, as leader of NASA's *Cassini* imaging team, she became a spokesperson for the spacecraft's power source.[3] Launched in 1997, *Cassini* spent seven years traveling to Saturn, where it began orbiting in 2004, studying the planet, its rings, and moons, and transmitting data back to Earth. The spacecraft was powered by a very special type of battery: a radioisotope thermoelectric generator (RTG)* . . . in other words, a nuclear battery.

In the months before the launch from Cape Canaveral, a movement led by physicist Michio Kaku had formed, opposing it. Supporters of the movement feared a rocket explosion at launch would spread radioactive dust all over southern Florida. They also claimed that if the spacecraft accidentally reentered Earth's atmosphere and burned up during a flyby, it would release enough radioactive material to kill everyone in the world. Which is just absurd.

---

* RTGs generate electricity by converting the heat released from the natural decay of radioactive materials into electric power using thermocouples.

Porco knew those fears were not based in reality. She told me, "I participated in interviews and debates and I wrote an op-ed to explain the facts... We, who have worked with scientific concepts all our lives, can easily forget how complex and confusing things can appear to the nonscientist. But we must continue with it, in this case especially, because the very health of the planet and the creatures living on it are at stake."[4]

My nuclear electricity journey came full circle when Porco agreed to come to the Save Diablo Canyon rally.

When the big day arrived, December 4, 2021, I had no idea what to expect. Hoff, Ortiz-Wines, and I had spent the week prior setting up folding tables on the lawns at Cal Poly State University, chatting with students and inviting them to the rally. We also stood for hours outside the plant in cold, damp weather, handing out flyers and hoping to inspire the workers to join us. Most of them had a "bless your heart" look on their faces and thought we were wasting our time. To be fair, almost everyone had the same attitude. One so-called energy expert in California even laughed at me, saying there was "no chance in hell" Diablo Canyon would stay open. Safe to say things were looking grim.

The morning of the rally, I arrived at the San Luis Obispo courthouse high on adrenaline. Just down the street, a group of eight women waited for me, sporting T-shirts that read "Save Clean Energy" and "Nuclear Energy Is Clean Energy." They held on to ropes attached to a thirty-three-foot blimp, which was supposed to represent a single metric ton of CO_2. If Diablo Canyon was closed and its output replaced with fossil fuels, an extra 7.2 *million* metric tons of CO_2 would be going into the atmosphere *every year*. I wanted people to grasp that scale. Imagine 7.2 million of those giant blimps filled with CO_2, hovering in the air and choking our atmosphere.

I joined the group, and we marched down the street with the blimp floating above us. Leading the way, fist in the air, I watched cars come to a halt as we chanted, "Save the plant. Save the planet," and, "We're on a mission, to stop all emissions." Together, we made our way to the courthouse, staging what would become the largest pronuclear rally in U.S. history. Not going to lie, the bar was pretty low here, but it was the largest one.

Supporters came from the local community, neighboring towns, and even as far as New York City. At the courthouse, a DJ played Dua Lipa, kids held colorful signs, and activists spoke with passion. In other words, the vibes were immaculate. Hoff and Zaitz spoke together, sharing their journeys from fearing nuclear electricity to becoming fierce advocates. During her speech, Porco appealed to the antinuclear audience, saying, "I am asking you, for the sake of your future, and the future of your progeny, please reconsider."

We also heard from my friend, the legendary musician Grimes, who in a surreal video asked politicians to reconsider the closure of Diablo Canyon. She described the problem of climate change as a "crisis mode" that required all the tools available to us.

A big announcement closed the event. My nonprofit, Save Clean Energy, had organized a letter urging Governor Newsom to delay the plant's closure. It featured signatures from seventy-nine of the top scientists and energy experts across the country, including former secretary of energy Steven Chu, who had served in Barack Obama's administration. Our case had been strengthened further by the MIT and Stanford study, which had dropped a few days prior. The study concluded that keeping the plant open through 2050 would save California up to $21 billion in power grid costs—plus preserve ninety thousand acres of land that would otherwise get covered in solar panels.

By September 1, 2022, I found myself in new territory—both mentally and physically. With a growing pregnancy bump pressing against my desk, I was glued to a Zoom meeting of the California legislature. My efforts weren't just about fighting for the planet anymore; they were about the future I was bringing my child into. The stakes felt so much higher. As the votes rolled in one after another, I was equally shocked and relieved. It quickly became clear there was overwhelming support. In the end, lawmakers voted 67–3 to delay Diablo Canyon's closure. At the time of writing this book, the plant is on track to stay open until 2030, and PG&E is actively pursuing a twenty-year license extension from the Nuclear Regulatory Commission. For the first time in years, I felt hopeful and excited. I was ready to welcome my baby into a world *I* had helped improve.

2021 Save Diablo Canyon rally in San Luis Obispo, California

This victory was a huge milestone, signaling a drastic shift in perspective. If nuclear electricity can make a comeback in California—the literal birthplace of the antinuclear movement—it can make it anywhere. Our win proved that even in the most staunchly antinuclear places, politicians and citizens are starting to realize that it will be impossible to move away from dirty fuels without nuclear.

The most amazing part of this story is that there was barely any outrage. There was no mass panic, no riots in tie-dye. While Friends of the Earth and some other antinuclear orgs weren't exactly thrilled, the general response was a resounding "meh." A 2021 poll showed that 58 percent of California residents favored keeping the plant open, and in San Luis Obispo, support shot up to 74 percent of voters.

That's not surprising. Research shows that if you live next to a nuclear power plant, you are probably in favor of it.[5] A survey found that 91 percent of people living near U.S. nuclear plants have a favorable impression of them. Most have positive opinions

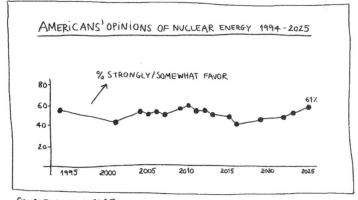

SOURCE: GALLUP, 2025

of nuclear electricity in general, and 78 percent support the addition of new reactors at their neighboring plant.[6]

Americans' support for nuclear is changing fast. According to Gallup's annual environment poll in March 2025, around 61 percent are in favor of nuclear electricity.[7] That might not sound groundbreaking, but it's a six-point jump from 2023, marking the highest public support in over a decade and just one point shy of the all-time record set back in 2010.

Around the world, support varies widely. Countries like Spain and Brazil are still extremely antinuclear.

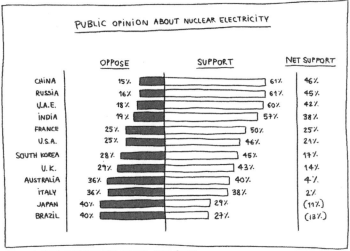

SOURCE: RADIANT ENERGY GROUP, NOV. 2023

The fact that nuclear electricity still has so many naysayers isn't all that surprising. There are a lot of powerful people working really, really hard to make you feel that way, and they've got decades of propaganda at their disposal. They've also got loads of cash. A recent analysis found that antinuclear NGOs spend fourteen times more money per year than pronuclear entities and rake in more than $2 billion annually.[8] But it's not working anymore.

People and countries everywhere are waking up to the potential of nuclear electricity. Even Japan has done a one-eighty. Within three years of the Fukushima accident in 2011, the government had closed all its reactors and fossil fuels went from making 55 percent to 75 percent of the country's electricity.

This shift resulted in higher electricity costs and carbon emissions. In response, Prime Minister Shinzo Abe advocated gradually reopening Japan's nuclear power plants. He set a new goal for nuclear to provide up to 22 percent of the country's electricity by 2030. As of 2024, ten out of the twenty-seven reactors that have applied for restarts are back online, with more on the way.

In Ukraine, where the Chernobyl disaster happened, support never faded. In 2021, nuclear provided more than half of the country's electricity. In 2024, Ukraine announced a partnership with an American company to build nine more reactors. If everything goes to plan, by 2035 the country could get more than 75 percent of its electricity from nuclear. Talk about a comeback arc.

But the ultimate proof of the vibe shift happened at COP28 in 2023, hosted by the United Arab Emirates. More than twenty-two countries made a bold pledge to triple global nuclear capacity by 2050—a level of support for nuclear never before seen at a global conference. In fact, nuclear used to be banned from many of them.

The United States, South Korea, Sweden, and a few others led the charge, admitting what's obvious to anyone who has more than two functioning brain cells: We're not hitting net zero by 2050 without doubling down on nuclear electricity.[9] The document they released summed it up (with my paraphrasing to keep things moving):

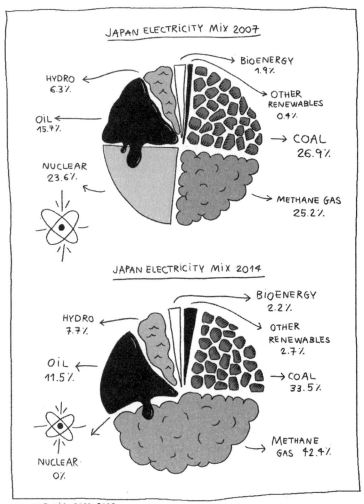

SOURCE: EIA, 2010, 2023

1. Nuclear electricity is already the second-largest source of clean, reliable power.
2. Research from the World Nuclear Association shows that if we want to reach global net-zero emissions by

2050, we need to triple our nuclear electricity capacity.
3. When the IPCC draws up models and plans for keeping climate change effects as minimal as possible, it generally finds that nuclear electricity's capacity needs to triple.
4. The International Energy Agency says nuclear electricity needs to more than double by 2050 to achieve net-zero emissions. IEA analysis also shows that decreasing nuclear electricity would make reaching net zero more difficult and costly.

In September 2024, over fourteen banks and financial institutions announced they'd step up with the financing needed to make this nuclear pledge happen. This was huge. For decades, many of these institutions wouldn't touch nuclear electricity with a ten-foot pole.

GENDER GAP

One quirky thing about our perception of nuclear electricity is the surprising gender gap. Across the globe, men are far more likely to support it than women. Experts say that even among trained physicists, dudes are far more likely to support nuclear than any of their peers. This may come down to how men and women perceive risk and societal power structures.

Jessica Lovering, engineer and cofounder of the Good Energy Collective, has argued that our worldviews can generally be mapped along one gradient: hierarchical versus egalitarian. Hierarchical societies are big on creating power structures and prioritizing respect for authority, with demonstrations of wealth and power given lots of weight. Egalitarian societies, on the other hand, lean toward equality and minimizing the distance between the most and least powerful. People fall all over this spectrum, of course, but most women tend to lean egalitarian, while most men skew hierarchical. Remember how back in the day nuclear became perceived as Big Government? Well, it seems like this image stuck.[10]

Lovering told the *Washington Examiner* that "one of the reasons people don't like nuclear is it's seen as hierarchical, pushed by big government and associations of the military-industrial complex."[11] Ironically, in the United States, the same utility companies that run nuclear plants also run big solar and wind farms. Unfortunately, they operate fossil fuel plants too.

Then there's the risk angle. Men tend to be more risk tolerant (that's why you'll see more bros doing the latest life-threatening Tik-Tok trend), while women tend to be more risk averse. But nuclear electricity isn't the risky business we have been led to believe it is. The perception of it as a high-risk, potentially high-reward technology has made it way more popular among the kind of guy who corners you at a party to talk about crypto. But that perception isn't based in reality. We have the opportunity to spread the gospel of an energy-abundant nuclear future with the radical girlies and Earth mothers of the world. Nuclear electricity is going to help us achieve equality, peace, and prosperity. That's the message we need to share if we're going to change the minds that matter.

Utility providers and private companies can make a big difference here. "You need to change business models and how to engage with communities, particularly at the very early stage when you're doing these projects, so that people feel ownership and that they're not being bulldozed about their energy choices," Lovering said. "If we can get that part right, we have a big chance at changing the way people feel about nuclear technology going forward."[12]

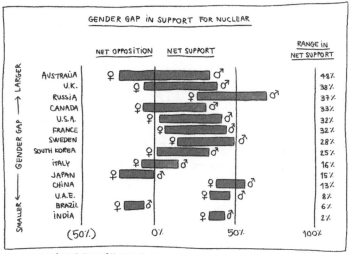

SOURCE: RADIANT ENERGY GROUP, 2023

Some research suggests that just hearing the word "nuclear" triggers our immune protection instincts.[13] Humans have evolved to have a healthy fear of anything that might make us sick. But just like our evolutionary protections against getting eaten by lions now leave us screaming our heads off at harmless horror movies, our adaptations against potential illness can misfire. For instance, some experts argue that racism and xenophobia boil down to an instinct to keep folks from outside our communities at a distance. That might have been a reasonable mentality when we lived in tiny nomadic tribes and newcomers could carry entire new worlds of germs inside them. These days it's a bad look at best.

All the stories about nuclear meltdowns, nuclear waste, and radioactive contamination make our well-meaning lizard brains perk up. Our bodies are giving us the ick to try to protect us, the same way they'd protect us from eating rotting meat or kissing somebody having a coughing fit. If that's true, then debunking myths about radiation and the risk of contamination could have a big impact.

THE NUCLEAR RENAISSANCE

By now, we have established that nuclear electricity got screwed by a decades-long smear campaign, even though it's one of the best sources of energy out there. We've also learned that, for a bunch of reasons, it has become expensive to build nuclear power plants in the West. Despite all that, governments are finally clocking that we ain't gonna make it without them. Welcome to the nuclear renaissance.

But we're at a crossroads. It's not like nuclear electricity won't be a part of the future. It's just a question of who will lead. China, for one, isn't messing around. According to the World Nuclear Association, China plans to build 150 reactors in fifteen years—a strategic push to combat air pollution, cut coal reliance, and secure its spot as the top dog in global energy. China is completing projects on time, on budget, and even exporting its tech, with plans to build thirty reactors in other countries by 2030. Russia is following suit with around twenty projects in the pipeline, including a new generation of microreactors.*

People often ask me what policy changes are needed to man-

---

* We will get to microreactors in a sec!

ifest our nuclear electricity future. In 2024, there wasn't a lack of pronuclear policy in the United States. The Department of Energy under President Biden rolled out report after report stressing the need to triple the nation's nuclear capacity. His administration set an ambitious target of 100 percent clean electricity by 2035—a goal that's pretty much fantasy without nuclear in the mix. To make clear it meant business, in 2022, the government passed the Inflation Reduction Act (IRA), creating incentives to keep current plants running, build new ones, and train the workforce needed to guarantee this nuclear future.

Disclaimer: As I was finishing this book, Donald J. Trump was inaugurated as the forty-seventh president of the United States. While President Trump has voiced his support for nuclear, it's unclear what will happen to the IRA and other incentives under his administration.

One thing is for certain: Trump's government is going all in on AI, as highlighted in his announcement of the Stargate Project, an American-based artificial intelligence collaboration formed by OpenAI, SoftBank, Oracle, and the investment company MGX. The project aims to invest $500 billion in four years to build new AI infrastructure and it will need an insane amount of electricity. In May of 2024, *The Telegraph* reported that an analyst at Morgan Stanley speculated that the project would be powered by several nuclear plants.[14]

It's not just the American government that's cheering for a little nuclear comeback. As we saw in the previous chapter, France had a bit of a dark-age moment in 2015 when it promised to cut nuclear's share from 70 percent to 50 percent of all electricity. President Macron had a change of heart and, in 2023, announced a fresh game plan: Keep all current plants open and build six more.

Meanwhile, Poland—still getting over 70 percent of its electricity from fossil fuels as of 2023—is planning to build eight reactors and invest a whole bunch of money into becoming a nuclear electricity hotspot.[15] Ghana is getting financial help from the United States to boost nuclear electricity education and development.[16] Even Italy, which shut down its last nuclear power plants after Chernobyl, is hopping back on the nuclear train.

People are buzzing with excitement about nuclear electricity *all over the world*.* But I am worried that this renaissance could very easily flop in the West.

Electricity providers in the United States are in a weird spot. They *know* they need to build more reactors—that's obvious. But they have no idea *what* to build. The United States just wrapped up its first nuclear reactor build in thirty years at the Vogtle Electric Generating Plant in Georgia.[17] The process wasn't a huge success, as I mentioned in the previous chapter. For comparison, China is out there, casually popping reactors of the same size in about four years and for $3 billion each.

Georgia taxpayers are understandably pissed off about footing the bill for delays and cost overruns, and "experts" (with vested interest in other forms of energy) have made lots of noise about how this power will cost way more than planned.[18] On the bright side, over half of the state's electricity is now clean. Call me an optimist, but I think that forty years from now, when Vogtle is still churning out clean energy and boosting the economy, folks will be glad it's there. In fact, some industries are already flocking to Georgia to take advantage of the plant's *reliable* and *clean* power.

Nonetheless, I get why utility companies look at Vogtle and think "Hard pass!" but the only way out is through. Experts point out that over the course of ten years, costs of wind and solar had

---

* Except for, of course, ze Germans.

plummeted by 70 percent and 89 percent respectively.[19] We're not getting to that point with nuclear unless we keep moving the ball forward. The Vogtle project may have been rough, but it's given us the experience and trained workforce we need to keep building. Repeating the process time and again is what drives costs. Get this: Costs went down about 30 percent from unit 3 to unit 4. It's game time, people. And by "people," I mean utilities and the nuclear industry.

We also need a few powerful private companies to make the bold leap into a nuclear future. Luckily, that's already happening.

Spurred by AI's immense power needs, Silicon Valley went nuclear. In March of 2024, Amazon Web Services (AWS) acquired a massive data center next to and powered by the Susquehanna nuclear plant in Pennsylvania. Later in that same year, AWS doubled down by signing an agreement with X-energy to make a big investment in the company and build its small high-temperature gas-cooled reactors (HTGR).*

Another company that is stepping up is Microsoft. In 2024, it signed a twenty-year deal with Constellation, one of the largest electricity producers in the United States. To meet Microsoft's needs, Constellation is planning on bringing the Three Mile Island nuclear power plant back online.†

Meanwhile, Oracle founder Larry Ellison grabbed headlines with his own bold move, announcing plans to power an AI data center with small modular reactors. The center will need over

---

* HTGRs are nuclear reactors that use helium gas as coolant to produce high heat, which can be turned into electricity and even used in industries needing extreme heat, like hydrogen production.

† The Three Mile Island power plant had two reactors. One of them had an accident in 1979, as we talked about in chapter 7. Constellation will restart the plant's other reactor, which kept operating without any issues until 2019.

1 gigawatt of electricity, powered by three small reactors. During an earnings call, he quipped, "The location . . . already [has] building permits for three nuclear reactors . . . This is how crazy it's getting. This is what's going on."[20]

Google isn't missing out either. In 2024, it ordered six small modular reactors from California startup Kairos Power, with the first reactor set to go live in 2030.

You might be wondering why these tech giants are betting on nuclear. It's simple. They want to be on the leading edge of artificial intelligence, and they know they'll need plenty of juice in the years to come. Training and running AI models requires a ton of electricity. For example, answering a single question with ChatGPT uses ten times more energy than a typical Google search, according to David Porter of the Electric Power Research Institute. As of November 2024, AI accounted for 10 to 20 percent of energy used by U.S. data centers.[21]

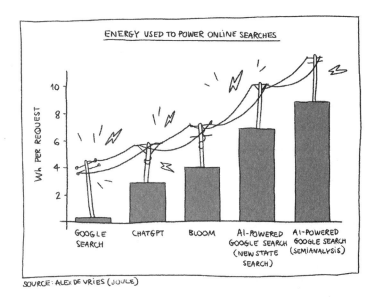

The International Energy Agency predicts that global electricity demand from data centers will more than double from 2022 to 2026, with AI playing a significant role in that increase.

There's just one problem. Many of those tech companies have seriously ambitious climate goals. Microsoft and Google, for example, want to be carbon negative by 2030, meaning they want to remove more carbon from the atmosphere than they put in. To hit that target, they'll need to ditch fossil fuels while at the same time keeping their data centers running. This is why nuclear reactors make perfect sense.

Nuclear offers something that no other source of energy can—reliable, clean electricity that can be built almost anywhere. It's the only option that checks all the boxes, with low emissions, constant supply, and enough juice to fuel the AI revolution.

SMALL MODULAR REACTORS

In recent years, small modular reactors (SMRs) have become the cool kids on the nuclear block. The name is straightforward: They're reactors that are—you guessed it—small and modular. They're between 5 megawatts and 300 megawatts, which can power anywhere from 3,000 to 220,000 American homes. The goal is to build these smol boys in a way that's similar to an assembly line, with a few modules coming out of factories and then snapped together like LEGO blocks at the site. This design promises to make them cheaper and faster to build compared to the big reactors of the past.

But the idea of a tiny nuclear power plant is as old as nuclear itself. The United States started building four pioneering small reactors in 1947, kicking off a whole family tree of designs. They tried everything: organic-cooled,* molten salt–fueled,† liquid mercury–cooled,‡ gas-cooled, heavy water—you name it, they experimented with it!

The Navy has also used small reactors famously well to power submarines. This offered a *huge* strategic advantage over fossil-powered subs. Nuclear subs don't need oxygen for combustion, don't reveal their locations with exhaust, and can carry decades' worth of fuel. Modern American submarines are so efficient they can last the entire planned life of the ship without refueling. They're also incredibly safe. Since 1954, there hasn't been a single incident of reactor failure or radiation leak from a U.S. Navy submarine.

The Army, not to be outdone, went on a microreactor road trip, putting them on remote bases in Greenland, Antarctica, Alaska, and even trucking one up a mountain in Wyoming. They also made a tiny truck-mounted reactor that operated in the Idaho desert and a floating mobile one that powered part of the Panama Canal for about a decade. But in each case, the Army eventually went back to using diesel or oil instead. It was just cheaper. The truck-mounted reactor, for example, was quoted as ten times more expensive than oil.

So where did all the tiny reactors go? They got big. By the 1960s, it became clear that some types—especially light water reactors§—were outshining the rest. Vendors realized that if they supersized them, they could finally compete with coal on cost.

---

* A type of nuclear reactor that uses organic liquids, like certain oils, instead of water or gas to cool the reactor core.

† Unlike traditional reactors that use solid uranium fuel rods, molten salt reactors dissolve fuel (such as uranium or thorium) in liquid salt.

‡ I think you get it!

§ Uses good ol' normal water for cooling the core and moderating the reactions.

And so, we went from reactors that could power a couple thousand homes to beefy ones that could power over a million. Fast-forward through that depressing slump (when the U.S. stopped building new reactors) to 2002, when we started hearing murmurs of a nuclear renaissance emerging. With climate change, oil shortages, and energy independence all over the news, nuclear engineering enrollments went up and orders for big reactors started rolling in. It was an exciting time. But then, I kid you not, four major things happened at once:

1. Fracking and horizontal drilling came along, making natural gas suuuper cheap. The U.S. went from importing gas to exporting it to the extreme. Coal and nuclear plants started shutting down in merchant markets where they simply could not compete with a giant jet engine bolted to the ground and hooked up to a generator powered by three guys and a pipeline.
2. Wind and solar marginal costs fell to one tenth of what they had been in the span of a decade.
3. Fukushima Daiichi.
4. Vogtle made us realize the U.S. was bad at managing reactor megaprojects.

Most of the big reactors ordered were canceled. It was impossible to find any utility willing to take on large nuclear projects. By a process of elimination, reactor vendors and promoters dusted off the concept of small and microreactors.

The idea came back in full force in the 2020s, with over seventy startups worldwide promising to make nuclear cheap again. These designs can get quite wild. Some claim they'll run on nuclear waste, others will be using meltdown-proof fuel (even if you tried, you couldn't make it melt down). Companies like X-energy have reactors that produce ultrahot steam for industrial uses. Whatever you need, there's a small modular reactor company out there claiming they've got it covered.

It's anyone's guess which of these will succeed, and the majority of these companies will probably fail, but here's the lay of the land:

Despite all the buzz, in 2024 there were only four operating small modular reactors in the world. Two high-temperature, gas-cooled reactors in China and two floating units in Russia. In the United States, progress is . . . let's just say more leisurely. In August 2024, Kairos Power received approval from the Nuclear Regulatory Commission to construct a demonstration 50 MWe reactor. TerraPower, an SMR company backed by Bill Gates, broke ground in Wyoming in 2024 on a 345 MWe sodium-cooled reactor that could make enough juice for about 284,000 homes. Oklo is working on the Aurora design, which can go up to 75 MWe, and Westinghouse is pushing forward with its eVinci 5 MWe microreactor, both planning demos at Idaho National Laboratory. Last Energy and Aalo Atomics, both based in Texas, are designing a 20 MWe and a 10 MWe respectively.

Old-school vendors, like GE Hitachi, are also in the mix, offering medium-ish (smedium?) reactors. These are essentially downsized versions of the tried-and-true reactors from the last century.

The bad news is that we also already know they're expensive. That's where companies like Microsoft, Amazon, Google, or Meta can make a difference. No matter what, the first few of any new kind of reactor are going to be a bit boondoggly and they're not going to save the companies that build them any money on energy costs. By going big on nuclear, these companies can not only fill their massive energy needs and meet their emissions goals, but they can also make reactors cheaper for everyone else. After all, history tells us that the more we build, the cheaper it gets. With time spent working out the design kinks, we'll eventually get better at making smaller reactors.

That won't just help make super high-tech, super tiny reactors to power individual buildings (though I hope we'll get to that point soon). By going small, we can try out a variety of designs without sinking endless time and money. Back in the fifties, the government funded all those little guys, and we learned everything we needed to know to build the big reactors we came to rely on. If we want to keep improving the way our large-scale reactors work, one way to go about it is to throw a bunch of small reactors at the wall and see what sticks. Maybe Microsoft can foot the bill this time. If they want Earth to keep looking like it does in that iconic screen saver of theirs, they'd better keep supporting nuclear.

Of course, there's always the South Korean and French approach—just pick one proven design, rally behind it, and streamline production. Perhaps tech companies can provide the money and space for nuclear engineers who want to play in that sandbox, while the biggest effort—and funds—goes toward designs we know we can rely on.

TL;DR

People around the world are waking up, shedding their outdated biases, and turning toward nuclear electricity. The nuclear renaissance is fully under way. It's being led by young people who are fed up with boomer environmentalism. This revival is being fueled and funded by an unlikely hero: artificial intelligence. We're witnessing the birth of AI and its insatiable energy needs that will most likely just keep growing. The companies behind it are betting big on nuclear electricity to meet those needs and are even helping us experiment with new reactor designs. In other words, the robots might save us. Not by overthrowing humanity, but by finally giving us a reason to stop being weird about nuclear electricity.

10: MANIFESTING A RAD FUTURE

Remember Otto Hahn and Fritz Strassmann's night at the lab? The night when they were messing around with a uranium sample and accidentally kick-started the nuclear age? Imagine if that fateful night hadn't gone down in Nazi Germany of all places, but instead had happened more recently.

Hard to know for sure, but I believe the world would be celebrating this groundbreaking discovery. We would see nuclear for what it is: a superior energy source that can solve climate change and energy inequality. We'd rush to develop nuclear reactors and usher in a radiant post–fossil fuel age. No oil wars. No smog. Just an abundance of clean energy. That is the promise of the atom.

But alas, that pivotal discovery took place back in 1938 while humanity was in one of her villain eras. So instead of a clean energy revolution, we got bombs. And hey, maybe that would have always been the case, humans do seem to have a knack for self-sabotage. The promise of the atom turned out to be a lot more complicated. Its powerful potential energy could be used for incredibly good or devastatingly bad ends, depending on who wields it or when.

The shock from the bombings of Hiroshima and Nagasaki, combined with the anxiety and fear that gripped the planet during the Cold War, traumatized an entire generation. It'd be like if the world was introduced to artificial intelligence by way of raging murderous robots trashing San Francisco, instead of innocent-sounding chatbots.

Nuclear electricity burst onto the scene with possibly the worst PR campaign of all time. As a result, we've spent decades grappling with the immense power—and immense responsibility—of the godlike technology we unleashed. All things considered, we've done a decent job. Between 1960 and 2024, the number of nuclear reactors around the world skyrocketed from 15 to 440. On the contrary, the number of nuclear weapons, which peaked at 70,300 in the 1980s, dropped to around 12,000 in 2024. This proves we can harness the promise of the atom for good, without destroying ourselves.

Despite that progress, nuclear's scary origin story cast a long shadow, holding humanity back from fully embracing its potential. That fear left us shackled to fossil fuels, filling our lungs with dirty air while pumping the atmosphere full of heat-trapping gases. Now, we're stuck on a warming planet plagued with rising sea levels, extreme weather, and raging wildfires. To add insult to injury, we still don't even have flying cars.

Our radical clean energy vision hasn't materialized just yet, but it's not too late. In fact, those years of misguided hostility toward nuclear electricity set the stage to develop it now in a much smarter and safer way—that's if we can get our act together.

Despite my annoyance at the OG environmental movement's crusade against nuclear electricity, their overcaution has ironically put us in a great position. Sure, they lumped weapons and energy together, convincing most people that anything with the word "nuclear" is inherently bad. (Which is like saying that electricity is bad because electric chairs are bad.) At the same time, they also pushed for higher safety practices and standards than you'll find in *any* other industry.

Nuclear electricity now ranks as one of the safest forms of energy, thanks in part to those vocal hippies. But it's time for them, and the ones who followed in their footsteps, to upgrade their software and update their worldview based on facts. The data doesn't just say nuclear is safe, it proves it's the energy source with the smallest environmental footprint. From the minimal materials required to build and operate power plants to how the waste is handled, nuclear rises above all those other bitches. Nuclear is radiant. Nuclear is radical. Nuclear is the missing ingredient to manifest a truly rad future.

In this future, the possibilities are as boundless as the power of the atom itself as we unleash energy to power technologies that make our lives better without harming the planet, polluting the air, or draining resources. Our greatest innovations can be supercharged, driving smarter work, faster scientific breakthroughs, and transformative healthcare advancements. AI supercomputers, advanced recycling, terraforming deserts and frozen wastelands, even exploring other planets can all become within reach. With energy abundance at our fingertips, the focus shifts from surviving to living our best life.

It's wild to think back to the moment I looked in the mirror and decided to become a nuclear energy influencer. It turns out you don't need a PhD to advocate for science. You don't need to be a policy wonk to push for change. You just need conviction, some audacity, and a mildly unhinged obsession with correcting people online. It can start small. Trust me, I had no idea my efforts would end up helping save a *whole nuclear power plant*. All I knew was that I wanted to share what I'd learned about this topic with others.

Despite all the chaos, all the setbacks, all the nights spent yelling into the void of the internet, I'm more optimistic than ever about our ability to tackle the climate crisis. We've got solutions. We've got momentum. And we've got proof that, when we keep at it, progress isn't just possible—it's inevitable. And no, my optimism isn't coming from some sparkly, rose-colored delusion about how great everything is. It comes from something much more grounded: knowing that I can *do* something about the future. And so can you. But first, we need to embrace a whole new way of thinking.

To have a great future, we have to think humans are great. Not because I have some weird obsession with our species (which I get sounds sus considering I am a part of it), but because you can't create positive change while believing in fundamental human worthlessness. It's obvious when you think about it, but the path to progress demands we see ourselves as worthy of that progress. This should also include other animals and the environment, as humans can't flourish if the rest of the world is completely out of whack. The human-hating mentality that has dominated our worldview for decades needs to go. It's not edgy or deep—it's just a fast-track to apathy and inaction. Instead, let's embrace what makes us cool: our ability to think, to care, to adapt. We have the chance, maybe the *responsibility*, to be conscious custodians of this weird little rock we call home.

We also need to develop superhuman problem-solving skills, which can only come from accepting problems as an inevitable part of life. If we don't, we'll give up the moment our solutions create new challenges, which they always do. For example, in the book we talked about how solar panels, wind turbines, and batteries need lots of raw materials and take up a bunch of land. We also learned about how nuclear electricity creates radioactive waste and can be expensive. Everything has downsides, but we can't let that stop progress. Switching to clean energy will require *far* less resources than maintaining our addiction to fossil fuels. Nuclear waste in dry casks has never hurt anyone or contaminated the environment, while burning fossil fuels creates loads of deadly toxic waste. The best we can do is keep pushing for incremental progress. Yeah, I know "incremental" doesn't exactly slap on a protest sign. It's not thrilling, it's not sexy. But it works. It is the only way to stop us from making "perfect" the enemy of "good," which is not just a buzzkill. In this case, it's also killing us.

HOW YOU CAN HELP

By now, you're hopefully convinced that there's no future without nuclear electricity. You're fired up and ready to make a rad future happen. Sadly, you can't go splurge on a personal mini nuclear reactor just yet. A lot of people write me asking what they can do, so I'll lay out different actions you can take—from "bare minimum" to "high effort."

BARE MINIMUM

HIGH EFFORT

BARE MINIMUM

Talk about it. The easiest thing you can do to manifest a nuclear future is to just . . . talk about it. Seriously, it works. Start with your friends—maybe those who are also concerned about climate change. From my experience, people don't know *anything* about nuclear, which means you don't need to dive straight into a pronuclear rant. Start by just asking questions to spark their curiosity. Here are a few that always work for me:

- Did you know that nuclear is the largest source of clean energy in the United States?
- Or know that nuclear is the second-largest source of clean energy in the world?
- Did you know that 70 percent of France's electricity comes from nuclear?
- And that nuclear waste stored in dry casks has never hurt anybody or harmed the environment?
- Two billion years ago, there were natural nuclear reactors on Earth. Isn't that crazy?

Most people haven't heard those points before—and if they're curious enough, they'll want to learn more. This opens a door to get them interested in learning about the topic on their own. As a great bonus, these prompts are also perfect for escaping the hellhole of small talk during networking events or long lines. Who wants to talk about the weather or people's summer plans when you could be talking about all-natural ancient nuclear reactors instead? You're about to become the most interesting person in line at the post office.

Social media is another powerful tool. If you see a big account posting misinformation about nuclear electricity, write a comment calling them out. Repost interesting facts and opinions too.

If you run a social media page that shares information on topics related to climate change, you can take some small steps to help rehab nuclear's image. Start by swapping the popular but useless phrase "renewable energy" with "clean energy." Anytime you share content related to clean energy, make sure to include nuclear front and center. Even something as small as tweaking your graphics can have a big impact. Some organizations use red or gray to symbolize nuclear electricity in their charts while using green or blue to symbolize other clean energy sources like solar and wind. This sends a subtle but insidious message: Nuclear electricity is dirty. It's dangerous. Deprogram your friends and followers by putting nuclear at the top of your clean energy lists and using lush green colors.

SLIGHTLY HIGHER EFFORT
Do some digging. Okay, now we're getting into stuff that takes a little more effort. Like voting. Make sure to always check your candidate's stance and record on nuclear electricity. What you will find is that some who claim to have climate change at the center of their platform are oddly silent on the issue. Politicians will rarely share their views on a topic they think is too controversial for voters. If your favorite candidate is quiet on nuclear or says vague things like "I want to invest in renewable energy," reach out to their office and ask where they stand on nuclear electricity specifically. Maybe even share some figures on how support for it is increasing.

If they're openly and ideologically antinuclear, write or call to let them know it's unfortunate they take this position, considering we need nuclear electricity to get off fossil fuels. Do the same for utilities: Find out which company provides your electricity and get in touch with them to ask what steps they are taking to get off fossil fuels and build nuclear power plants.

I know this sounds small, but if a candidate or a utility receives even one of these messages a day, it can make a difference. Humans are very bad at grasping scale. Just ask anyone who's been mobbed on X. Receiving two hundred negative responses on a post might make it seem like the whole world is mad at you. Of course, two hundred people represent about 0.000000025 percent of the global population. That's ... far from the whole world. You can harness this brain quirk for good by making companies and politicians feel like being pronuclear electricity is hot.

You should also find out what the environmental nonprofits you support are up to. A few of them take a very loud antinuclear stance, Greenpeace being the biggest offender. If you go on their website and search for "nuclear energy," you'll see articles saying things like "Nuclear energy has no place in a safe, clean, sustainable future." If you're currently donating to them, consider stopping. Write to let them know why: Their antinuclear electricity activism is anti-science and bad for the planet. The goal is to show these influential groups that public perception is changing and people will no longer allow organizations to actively harm the planet with their outdated views. Start donating to organizations that *do* support nuclear electricity, or at least aren't actively fighting against it. For a list of those organizations, go to www.isodope.com/radfuture.

HIGH EFFORT

Join the movement. If you're fully in and feel a pull to dedicate some of your time to manifesting a rad future, join an existing pronuclear organization and become a volunteer. You can help by writing opinion pieces, circulating petitions, and attending demonstrations. You can get plugged in to local events to help educate your friends, neighbors, and elected officials. Whatever your passions and skills are, there's a way for you to use them. You just need to get out there and connect with this growing network of rad future visionaries. I'll also post a list of these organizations at www.isodope.com/radfuture.

I started this journey, and this book, with the wildfires raging in the Amazon and Australia. Hate to be the bearer of apocalyptic symmetry, but I'm ending it with wildfires too.

In January 2025, Los Angeles became hell on earth as uncontrollable flames ripped through the city. The Palisades Fire, the Eaton Fire, and other fires devoured over 57,000 acres, destroying more than 12,000 structures. At least 200,000 people had to flee their homes.

This nightmare resulted from a deadly combination of incompetence, bad luck, and a warming planet. Poor vegetation management left dry brush piled up waiting to catch fire, while record-breaking Santa Ana winds turned every ember into a flamethrower. Add in the fact that vegetation is parched from a +3.6°F bump in fire-season temperatures, and you've got the perfect recipe for the worst wildfire in California's history.

You've made it this far in the book, so I'm sure you already know climate change is behind that increase in temperature. Which brings me to my final point: Our planet is literally on fire, and we are *still* wasting time debating whether nuclear electricity should even be considered part of the solution.

The fear, misinformation, and historical baggage around nuclear electricity are tired and no longer serve us. It's a proven, scalable, and clean energy source that can replace fossil fuels while keeping the lights on. History has shown it works. Science confirms it's safe. And the urgency of our climate crisis makes it clear that ignoring it is not an option anymore. Today's fires are just a preview of what's coming if we keep screwing around.

Let's be brave enough to move past outdated narratives and ideological hang-ups. Let's be smart enough to embrace solutions that work, even if they're not perfect. Because while we debate . . . the world burns.

ACKNOWLEDGMENTS

They say raising a child takes a village, and apparently so does writing a book. This book truly wouldn't exist without a lot of people and there's no way I will do justice to them all.

It all started with Carolyn Porco, the legendary planetary scientist whose single X post (back when it was Twitter) sparked my obsession with nuclear electricity. Who would have thought that a random scroll would lead to this moment? I'm endlessly grateful for her encouragement, for joining the campaign to save Diablo Canyon, and for becoming a friend along the way.

Then there's Nick Touran, nuclear engineer and the first industry insider I pitched my idea of becoming a "nuclear energy influencer" to. I'm still shocked he didn't laugh me off the phone. We met just before COVID flipped the world upside down, and since then, he's been my nuclear encyclopedia. ISODOPE wouldn't exist if it weren't for the immense work he's put in for decades, compiling all the data you could possibly need about nuclear on his website whatisnuclear.com. Plus he has patiently responded to all my questions and requests throughout the years.

I've been lucky to lean on other experts too: James Krellenstein, who reviewed this book even from the trenches of launching a critical nuclear start-up. Mark Nelson, a tireless super connector who's been there since the Diablo Canyon campaign and always has my back. Staffan Qvist, who reviewed the book while holding a newborn baby. And to the sharp minds in the Arcane Nuclear Beefs group chat: keep the beefs going. Also have to tip my hat to Simon Holmes à Court, who's been an excellent sparring partner and forced me to level up my game.

On the book side, huge props to Helen Healey-Cunningham and Megan Wenerstrom, who didn't lose faith in me, even after I changed the entire book structure no less than twenty times. Lauren Hall and Rachel Feltman supported this vision and helped me focus when I had no idea what I was doing. Simon Sinek's blunt feedback ("This chapter's boring") made this a much more fun read for everyone. I am also lucky to call him a great friend.

Talking about friends, I cannot possibly write the name of all those who helped me in big or small ways. From edits to words of encouragement to just being in my life. Lydia Kives, Liv Boeree, Ola Abrams, Rosana Bischoff, Tandice Urban, Christiana Musk, C, Igor Kurganov, Tim Urban, Michael Kives, Bryn Mooser, Annaka Harris, GP, Jessica Seinfeld, Veronica Grazer, Brian Grazer, KP, Orly, Sophie Weld, Terry Lee, Jordan Grenier, and many, many others. A special shout-out to Nicole Nosek, who helped me launch the Save Diablo Canyon campaign. She rolled up her sleeves gathering signatures for our letter to Governor Newsom to rethink the plant's closure. All that while super pregnant and busy with her own advocacy work.

While we're on the topic of Diablo Canyon, I was continuously inspired by Heather Hoff from Mothers for Nuclear, who never gave up the fight to save the plant. Between her, Paris Ortiz-Wines, and Mark Nelson, we somehow transformed Zoom calls with three rogue advocates into a campaign that helped achieve the impossible. Big thanks to Eric Meyer from Generation Atomic and Ryan Pickering for joining the fight and putting so much effort into it. But the campaign wouldn't have been successful without Armond Cohen from Clean Air Task Force and Jacopo Buongiorno from MIT. These two deserve a medal for doing the right thing, even though the chances of success were slim.

I don't know if publicly thanking your therapist is a thing. But it felt wrong not to thank my therapist, Dr. JK, for keeping me sane through the writing process and helping me flesh out some ideas.

My late grandmother Dida's determination and passion for knowledge inspire me every day, and I can only hope to have that impact on my own grandchildren. My parents, Denise and Paulo, cheered me on from day one, even when my nuclear video rants left them baffled. Same for my in-laws, Joe and Eileen.

Finally, to my husband, Joe, for being my number one cheerleader and enabling me to pursue my wildest dreams. He and our son make me want to be a better human and fill my life with love and happiness.

GLOSSARY

ATOM: The basic building blocks that come together to create all the matter in the universe. Each atom contains smaller particles called subatomic particles. The center of the atom is called the nucleus, which contains positively charged protons and uncharged neutrons. The nucleus is surrounded by a probability cloud of negatively charged electrons.

BIOMASS: Organic matter that can be used as fuel, including stuff like wood, crop residues (straw, wheat husks), animal poop, and used cooking oil. It's considered renewable because plants can be regrown, and organic waste is continuously produced. While it sounds good in theory, burning biomass still releases significant air pollutants and greenhouse gases.

CAPACITY FACTOR: The ratio of actual energy output over a period compared to the maximum possible output if the facility operated at full power continuously. If you have a solar farm in, let's say, Germany, your solar farm might only produce its potential electricity output 10 percent of the time (because it is not very sunny up there). That means you would have a capacity factor of 10 percent. (Lame.)

Glossary

CARBON FOOTPRINT: The total amount of greenhouse gas emissions caused by an activity, person, organization, or country, usually expressed in equivalent metric tons of carbon dioxide (CO_2e). This includes not only CO_2 but other greenhouse gases, like methane. The concept was popularized in the early 2000s as a result of a marketing campaign by BP, a British oil and gas company.

CHAIN REACTION: A series of events where each event triggers the next one. In nuclear physics, a chain reaction happens when neutrons released from splitting atoms cause additional atoms to split, releasing more neutrons and creating a continuous cycle of fission.

CONTAINMENT DOME: A large dome-shaped structure built around a nuclear reactor, typically made of steel-reinforced concrete up to five feet thick. It serves as an additional barrier between the environment and radiation, in case things go wrong and there's an accident.

DEGROWTH: This is an economic theory and social movement that aims to shrink economies. Basically, these folks think that having reliable electricity and modern comforts is the problem, so their solution is to deliberately scale back production. It's like suggesting we solve world hunger by telling everyone to eat less.

DELULU: Internet slang for "delusional." Here, I'll use it in a sentence: "People who think nuclear power plants are more dangerous than coal power plants are delulu."

DRY CASK STORAGE: A method of storing spent nuclear fuel rods (waste) from nuclear power plants in special containers called "dry casks," which have walls more than a foot thick. Countries around the world have been successfully storing spent nuclear fuel for decades, without a single incident.

ELECTRON: A negatively charged fundamental particle (can't be broken down into smaller components) that exists in a probability cloud around the nucleus of an atom. Electrons can move between atoms—this movement of charge generates what we call "electrical energy."

ENERGY: Simply defined, this is the ability to do work. It comes in many forms like thermal (heat), electromagnetic (light), electrical, chemical (as in batteries), and kinetic (motion). Following the law of conservation of energy, energy cannot be created or destroyed, only transformed between different forms. For example when you make toast, electrical energy from an outlet transforms into thermal energy in a toaster, transfers to bread as heat, changes the bread's chemical energy through cooking, and finally converts to chemical energy in your body when eaten.

FISSION: Nuclear fission happens when a heavy atomic nucleus splits into two or more lighter nuclei (plural of nucleus), releasing neutrons and large amounts of energy in various forms including kinetic. Kinetic energy then creates more collisions, causing vibrating atoms that create heat.

FOSSIL FUELS: Technically, these are nonrenewable energy sources formed from dead ancient plants and animals that were buried underground over millions of years, subjected to high pressure and heat. Examples are coal, petroleum, and methane gas. For the purposes of this book, fossil fuels are air-polluting, greenhouse gas–releasing sources of fuel we need to get away from.

FUSION: Nuclear fusion happens when atomic nuclei combine to form a heavier nucleus, releasing enormous amounts of energy. In our sun, this happens when hydrogen nuclei fuse to form helium nuclei under conditions of extreme temperature and pressure.

Glossary

GREENHOUSE GAS: Gases in our atmosphere that allow sunlight to reach Earth but absorb some of the heat that would otherwise escape into space. This greenhouse effect is vital for life on Earth, maintaining temperatures warm enough for survival. The problem is that human activities (like burning fossil fuels at a massive scale) have rapidly increased the concentration of these gases in the atmosphere, causing too much warming. Common greenhouse gases include carbon dioxide (CO_2), methane (CH_4), and nitrous oxide (N_2O).

MANHATTAN PROJECT: A secret U.S. military program (1942–46) during World War II that developed and produced the first nuclear weapons. Led by physicist J. Robert Oppenheimer at the Los Alamos laboratory, the project culminated in the creation and use of two atomic bombs on Japan in 1945, marking the first and only use of nuclear weapons in warfare (so far).

MICROREACTOR: A reactor that produces under 20 megawatts of electricity. In America a reactor of that size would be able to power around 20,000 homes. For comparison, SMRs—or small modular reactors—produce between 20 and 300 megawatts.

MILLISIEVERT: A thousandth of a sievert. A sievert is a unit for measuring how much biological damage a dose of radiation is likely to cause to the human body, based on both the amount of energy absorbed and the type of radiation. Millisieverts just make it easier to talk about smaller, more everyday doses, like what you'd get from an X-ray or a long flight.

NEUTRON: An uncharged subatomic particle that helps keep the nucleus of an atom stable. When absorbed by certain heavy nuclei (like uranium-235), neutrons can cause nuclear fission.

NGOS: Nongovernmental organizations. These are nonprofit organizations that operate independently from the government. They are often involved in humanitarian and social causes.

OBVI: Short for "obviously." (Obviously.)

PROLIFERATION: The spread of nuclear weapons, nuclear technology, and nuclear materials to countries that are not recognized as "nuclear weapon states" by the Treaty on the Non-Proliferation of Nuclear Weapons, or NPT.

PROTON: A positively charged subatomic particle found in an atom's nucleus. The number of protons in an atom defines which element it is—for example, all carbon atoms have six protons and all uranium atoms have ninety-two protons.

RADIATION: Energy that travels through space or a material as waves or particles. This includes electromagnetic radiation (like light, X-rays, and gamma rays) and particle radiation (like alpha and beta particles).

RENEWABLES: A term used to describe sources of energy that can be replenished, like trees, poop, waste, sun, wind, and water. It is mostly used to exclude nuclear from the clean energy conversation and doesn't account for whether a source of energy emits air pollution or greenhouse gases. Let's get rid of it and use "clean energy" instead.

SPENT FUEL: Nuclear fuel that has been used in a reactor and is no longer efficient for power generation. Some people call it waste, but we don't because over 90 percent of it can be reprocessed and used to make electricity again.

THE ICK: Internet slang for something making you feel uncomfortable or disgusted.

TRITIUM: A radioactive isotope of hydrogen containing one proton and two neutrons. While it occurs naturally in small amounts in the upper atmosphere (created by cosmic rays), it can also be produced in nuclear reactors.

TURBINE: A machine with blades that is spun by flowing steam, air, water, or other fluids to convert kinetic energy into mechanical energy. When connected to a generator, this rotational motion is converted into electricity.

WATT: Unit that measures the rate of energy transfer or power consumption at a specific moment. Unlike watt-hours, which measure energy used over time, watts tell us how much power something needs to operate right now. Your typical LED household light bulb uses about 4 to 9 watts when it's on.

WATT-HOUR: Measurement of how much energy something consumes over time, calculated by multiplying power (watts) by hours. For example, if a 4-watt LED light bulb runs for 10 hours, it consumes 40 watt-hours of energy. You'll often see this measured as kilowatt-hours (1,000 watt-hours, shortened to kWh), megawatt-hours (1,000 kilowatt-hours, shortened to MWh), and gigawatt-hours (1,000 megawatt-hours, shortened to GWh). When measuring a country's energy use, it can go up to terawatt-hours (1,000 gigawatt-hours, shortened to TWh).

NOTES

CHAPTER 1: THE DAWN OF THE ATOMIC AGE
1. Naval Ravikant and David Deutsch, "David Deutsch: Knowledge Creation and the Human Race, Part 1," *Naval* (blog), February 11, 2023, https://nav.al/david-deutsch.

CHAPTER 2: HUMANS AND ENERGY
1. "Energy," Merriam-Webster.com, accessed January 28, 2025, https://www.merriam-webster.com/dictionary/energy.
2. Tim Urban, "Energy for Dummies," *Wait But Why* (blog), March 12, 2014, https://waitbutwhy.com/2014/03/energy-dummies.html.
3. Hannah Ritchie and Pablo Rosado, "Energy Mix," Our World in Data, updated January 2024, https://ourworldindata.org/energy-mix.
4. Meng Wang et al., "Association Between Long-term Exposure to Ambient Air Pollution and Change in Quantitatively Assessed Emphysema and Lung Function," *Journal of the American Medical Association* 322, no. 6 (2019), https://doi.org/10.1001/jama.2019.10255.
5. *Statistical Review of World Energy*, Energy Institute 2024, https://www.energyinst.org/statistical-review; "U.S. Energy Facts Explained," U.S. Energy Information Administration, updated September 2023, https://www.eia.gov/energyexplained/us-energy-facts.
6. *Monthly Energy Review April 2024* (U.S. Energy Information Administration, 2024), https://www.eia.gov/totalenergy/data/monthly/archive/00352404.pdf.

CHAPTER 3: FOSSIL FUELS' FINAL BOSS
1. Anne White, "Nuclear Energy," *MIT Climate Portal*, updated July 24, 2024, https://climate.mit.edu/explainers/nuclear-energy.
2. "Nuclear Power in the World Today," World Nuclear Association, updated September 2023, https://world-nuclear.org/information-library/current-and-future-generation/nuclear-power-in-the-world-today.aspx.
3. "Capacity Factors for Selected Energy Sources in the United States in 2023," Statista, April 2024, https://www.statista.com/statistics/183680/us-average-ca

pacity-factors-by-selected-energy-source-since-1998; "Electric Power Monthly: 6.7.A, 6.7.B, 6.7.C," U.S. Energy Information Administration, September 24, 2024, https://www.eia.gov/electricity/monthly.
4. Emma Derr, "Advanced Nuclear Fits the South Pole's Energy Needs," Nuclear Energy Institute, February 16, 2023, https://www.nei.org/news/2023/advanced-nuclear-fits-the-south-poles-energy-needs.
5. Julie Kozeracki et al., *Pathways to Commercial Liftoff: Advanced Nuclear*, U.S. Department of Energy, 2024, https://liftoff.energy.gov/wp-content/uploads/2024/10/LIFTOFF_DOE_Advanced-Nuclear_Updated-2.5.25.pdf.
6. Jane C. S. Long et al., "Clean Firm Power Is the Key to California's Carbon-Free Energy Future," *Issues in Science and Technology*, March 24, 2021, https://issues.org/california-decarbonizing-power-wind-solar-nuclear-gas.
7. "Nuclear Energy Factsheet," Pub. No. CSS11-15, Center for Sustainable Systems, University of Michigan, 2024, https://css.umich.edu/publications/factsheets/energy/nuclear-energy-factsheet.
8. Hannah Ritchie, "How Does the Land Use of Different Electricity Sources Compare?," Our World in Data, June 16, 2022, https://ourworldindata.org/land-use-per-energy-source.
9. Alex Breckel et al., *Growing the Grid: A Plan to Accelerate California's Clean Energy Transition* (Clean Air Task Force, 2022), https://cdn.catf.us/wp-content/uploads/2022/10/11081420/growing-grid-plan-accelerate-californias-clean-energy-transition.pdf.
10. Justin Aborn et al., "An Assessment of the Diablo Canyon Nuclear Plant for Zero-Carbon Electricity, Desalination, and Hydrogen Production," November 2021, https://drive.google.com/file/d/1RcWmKwqgzvIgIIh0BB2s5cA6ajuVJJzt/view.
11. Melissa Harris, Marisa Beck, and Ivetta Gerasimchuk, *The End of Coal: Ontario's Coal Phase-Out*, International Institute for Sustainable Development, 2015, https://www.iisd.org/system/files/publications/end-of-coal-ontario-coal-phase-out.pdf; "Provincial and Territorial Energy Profiles: Ontario," Canada Energy Regulator, updated June 2022, https://www.cer-rec.gc.ca/en/data-analysis/energy-markets/provincial-territorial-energy-profiles/provincial-territorial-energy-profiles-ontario.html; "Ontario's System-Wide Electricity Supply Mix: 2003 Data," Ontario Energy Board, October 26, 2005, https://www.oeb.ca/documents/electricity_mix_261005.pdf.
12. *Carbon Neutrality in the UNECE Region: Integrated Life-Cycle Assessment of Electricity Sources*, United Nations Economic Commission for Europe, 2022, https://unece.org/sites/default/files/2022-04/LCA_3_FINAL%20March%202022.pdf.
13. *Investigating Benefits and Challenges of Converting Retiring Coal Plants into Nuclear Plants*, Idaho National Laboratory, 2022, https://fuelcycleoptions.inl.gov/SiteAssets/SitePages/Home/C2N2022Report.pdf.
14. Paul Denholm et al., *Examining Supply-Side Options to Achieve 100% Clean Electricity by 2035*, National Renewable Energy Laboratory, 2022, https://www.nrel.gov/docs/fy22osti/81644.pdf.
15. OECD Nuclear Energy Agency and International Atomic Energy Agency, *Measuring Employment Generated by the Nuclear Power Sector*, 2022, https://www.oecd-nea.org/jcms/pl_14912/measuring-employment-generated-by-the-nuclear-power-sector?details=true.

16. Sean McGarvey, "Saving Nuclear Energy Plants Means Saving Jobs," *The Hill*, January 21, 2015, https://thehill.com/blogs/pundits-blog/energy-environment/241819-saving-nuclear-energy-plants-means-saving-jobs.
17. "The Paris Agreement," UNFCC, accessed October 2024, https://unfccc.int/process-and-meetings/the-paris-agreement.
18. IPCC, *Global Warming of 1.5°C: An IPCC Special Report on the Impacts of Global Warming of 1.5°C Above Pre-industrial Levels and Related Global Greenhouse Gas Emission Pathways, in the Context of Strengthening the Global Response to the Threat of Climate Change, Sustainable Development, and Efforts to Eradicate Poverty*, ed. Valérie Masson-Delmotte et al. (Cambridge University Press, 2018), https://doi.org/10.1017/9781009157940.
19. Roxana Bardan, "Temperatures Rising: NASA Confirms 2024 Warmest Year on Record," NASA, January 10, 2025, https://www.nasa.gov/news-release/temperatures-rising-nasa-confirms-2024-warmest-year-on-record/.
20. Laura Gil, "Meet Oklo, the Earth's Two-Billion-Year-Old Only Known Natural Nuclear Reactor," International Atomic Energy Agency, August 10, 2018, https://www.iaea.org/newscenter/news/meet-oklo-the-earths-two-billion-year-old-only-known-natural-nuclear-reactor.
21. Sam Kean, host, "The World's Only Natural Nuclear Reactor," *The Disappearing Spoon*, podcast, Science History Institute, June 1, 2023, 21 min., 48 sec., https://sciencehistory.org/stories/disappearing-pod/the-worlds-only-natural-nuclear-reactor/.
22. Sarah Tse, "The Two-Billion-Year-Old Nuclear Reactors of Gabon, Africa," *Science Explorer*, September 8, 2015, https://www.thescienceexplorer.com/the-two-billion-year-old-nuclear-reactors-of-gabon-africa-34.
23. Sam Kean, "The World's Only Natural Nuclear Reactor."
24. Rugile Paleviciute, "The Environmental Impact of the Denim Industry," *Make Fashion Better* (blog), March 12, 2023, https://www.makefashionbetter.com/blog/the-environmental-impact-of-the-denim-industry.
25. Levi Strauss & Co., "The Life Cycle of a Jean," 2015, https://levistrauss.com/wp-content/uploads/2015/03/Full-LCA-Results-Deck-FINAL.pdf.
26. Nenad Raisić, "Desalination of Sea Water Using Nuclear Heat," *Bulletin of the International Atomic Energy Agency* 19, no. 1 (1977), https://www.iaea.org/sites/default/files/publications/magazines/bulletin/bull19-1/19105982126.pdf.

CHAPTER 4: DISPELLING MYTHS

1. Justin Higginbottom, "Dam Collapse That China Kept Secret," *MCLC List* (blog), Ohio State University, February 18, 2019, https://u.osu.edu/mclc/2019/02/18/dam-collapse-that-china-kept-secret.
2. Scientific Committee on the Effects of Atomic Radiation, "Assessments of the Radiation Effects from the Chernobyl Nuclear Reactor Accident," United Nations, accessed October 2024, https://www.unscear.org/unscear/en/areas-of-work/chernobyl.html.
3. Vladimir Drozdovitch, "Radiation Exposure to the Thyroid After the Chernobyl Accident," *Frontiers in Endocrinology* 11 (January 4, 2021), https://doi.org/10.3389/fendo.2020.569041.

4. "The Real Story of the Chernobyl Divers," Sky History, accessed October 2024, https://www.history.co.uk/article/the-real-story-of-the-chernobyl-divers.
5. *Evaluation of Data on Thyroid Cancer in Regions Affected by the Chernobyl Accident*, United Nations Scientific Committee on the Effects of Atomic Radiation, 2018, https://www.unscear.org/docs/publications/2017/Chernobyl_WP_2017.pdf.
6. Hannah Ritchie, "What Was the Death Toll from Chernobyl and Fukushima?," Our World in Data, July 24, 2017, https://ourworldindata.org/what-was-the-death-toll-from-chernobyl-and-fukushima.
7. Alex Gabbard, "Coal Combustion: Nuclear Resource or Danger," *Oak Ridge National Laboratory Review* 26, no. 3/4 (1993), https://www.nrc.gov/docs/ML0636/ML063620175.pdf.
8. Mahir Aliyev, "How Chernobyl Has Become an Unexpected Haven for Wildlife," UN Environment Programme, September 16, 2020, https://www.unep.org/news-and-stories/story/how-chernobyl-has-become-unexpected-haven-wildlife.
9. Pushker A. Kharecha and James E. Hansen, "Prevented Mortality and Greenhouse Gas Emissions from Historical and Projected Nuclear Power," *Environmental Science & Technology* 47, no. 9 (2013): 4889–95, https://pubs.acs.org/doi/epdf/10.1021/es3051197?ref=article_openPDF.
10. "Doses in Our Daily Lives," U.S. Nuclear Regulatory Commission, updated April 26, 2022, https://www.nrc.gov/about-nrc/radiation/around-us/doses-daily-lives.html.
11. "Radiation, How Much Is Considered Safe for Humans?," *MIT News*, January 5, 1994, https://news.mit.edu/1994/safe-0105.
12. Randall Munroe, "Radiation," *XKCD* (blog), https://xkcd.com/radiation.
13. Aurélia Beigneux, "Parliamentary Question E-003567/2022: Radioactive Ash from Coal Power Plants," European Parliament, November 4, 2022, https://www.europarl.europa.eu/doceo/document/E-9-2022-003567_EN.html.
14. James Conca, "How HBO Got It Wrong on Chernobyl," *Nuclear Newswire*, American Nuclear Society, July 10, 2019, https://www.ans.org/news/article-2143/how-hbo-got-it-wrong-on-chernobyl.
15. Charles Mitchell, "Abortions Recommended for Some Soviet Women," United Press International, June 6, 1986, https://www.upi.com/Archives/1986/06/06/Abortions-recommended-for-some-Soviet-women/9353518414400.
16. Linda E. Ketchum, "Lessons of Chernobyl, Part II," *Journal of Nuclear Medicine* 28, no. 6 (1987): 933–38, https://jnm.snmjournals.org/content/jnumed/28/6/933.full.pdf.
17. D. Trichopoulos et al., "The Victims of Chernobyl in Greece: Induced Abortions After the Accident," *British Medical Journal (Clinical Research Ed.)* 295 (1987): 1100, https://doi.org/10.1136/bmj.295.6606.1100; L. B. Knudsen, "Legally-Induced Abortions in Denmark After Chernobyl," *Biomedicine & Pharmacotherapy* 45, no. 6 (1991): 229–31, https://doi.org/10.1016/0753-3322(91)90022-L.
18. William J. Broad, "Rise in Retarded Children Predicted from Chernobyl," *New York Times*, February 16, 1987, https://www.nytimes.com/1987/02/16/us/rise-in-retarted-children-predicted-from-chernobyl.html.
19. Drozdovitch, "Radiation Exposure."
20. "Chernobyl: The True Scale of the Accident," press release, United Nations, September 6, 2005, https://press.un.org/en/2005/dev2539.doc.htm.

Notes

21. Cordula Meyer and Shunichi Yamashita, "Studying the Fukushima Aftermath: 'People Are Suffering from Radiophobia,'" *Der Spiegel*, October 11, 2011, https://www.spiegel.de/international/world/studying-the-fukushima-aftermath-people-are-suffering-from-radiophobia-a-780810.html; James Conca, "Fukushima—Fear Is Still the Killer," *Forbes*, March 18, 2013, https://www.forbes.com/sites/jamesconca/2013/03/18/fukushima-fear-is-still-the-killer/?sh=70a6ba3c54c9.
22. Organisation for Economic Co-operation and Development, "Perceptions and Realities in Modern Uranium Mining," 2014, https://www.oecd-nea.org/upload/docs/application/pdf/2019-12/7063-mehium-es.pdf.
23. "Environmental Aspects of Uranium Mining," World Nuclear Association, April 10, 2017, https://world-nuclear.org/information-library/nuclear-fuel-cycle/mining-of-uranium/environmental-aspects-of-uranium-mining.
24. Ariel Gould, "Sustainable and Ethical Uranium Mining: Opportunities and Challenges," *Good Energy Collective Policy Report*, August 31, 2022, https://www.goodenergycollective.org/policy/sustainable-and-ethical-uranium-mining-opportunities-and-challenges.
25. Jonathan Nez and Myron Lizer, "On Uranium Mining: Contamination and Criticality and H.R. 3405, the Uranium Classification Act of 2019," written statement of the Navajo Nation prepared for the House Committee on Natural Resources Subcommittee on Energy and Mineral Resources, July 12, 2019, https://www.congress.gov/116/meeting/house/109694/documents/HHRG-116-II06-20190625-SD013.pdf.
26. Laurel Morales, "For the Navajo Nation, Uranium Mining's Deadly Legacy Lingers," *Weekend Edition Sunday,* NPR, April 10, 2016, https://www.npr.org/sections/health-shots/2016/04/10/473547227/for-the-navajo-nation-uranium-minings-deadly-legacy-lingers.
27. R. J. Roscoe et al., "Mortality Among Navajo Uranium Miners," *American Journal of Public Health* 85, no. 4 (1995): 535–40, https://www.ncbi.nlm.nih.gov/pmc/articles/PMC1615135/.
28. Maria Harvey, "The Impacts of Mining on Indigenous Populations," Oregon State University, May 25, 2020, https://storymaps.arcgis.com/stories/52b944e9d1874f05ad595e330e6d7994; Nicole Greenfield, "Lithium Mining Is Leaving Chile's Indigenous Communities High and Dry (Literally)," NRDC, April 26, 2022, https://www.nrdc.org/stories/lithium-mining-leaving-chiles-indigenous-communities-high-and-dry-literally.
29. Sammy Witchalls, "The Environmental Problems Caused by Mining," Earth.org, April 3, 2022, https://earth.org/environmental-problems-caused-by-mining.
30. Amnesty International, "'This Is What We Die For': Human Rights Abuses in the Democratic Republic of the Congo Power the Global Trade in Cobalt," 2016, https://www.amnesty.org/en/wp-content/uploads/2021/05/AFR6231832016ENGLISH.pdf.
31. Charlotte Davey, "The Environmental Impacts of Cobalt Mining in Congo," Earth.org, March 28, 2023, https://earth.org/cobalt-mining-in-congo.
32. Madhumitha Jaganmohan, "Cobalt Demand Worldwide from 2010 to 2025," Statista, April 25, 2024, https://www.statista.com/statistics/875808/cobalt-demand-worldwide.

33. Hannah Ritchie, "Mining Quantities for Low-Carbon Energy Is Hundreds to Thousands of Times Lower Than Mining for Fossil Fuels," *Sustainability by Numbers* (blog), January 18, 2023, https://www.sustainabilitybynumbers.com/p/mining-low-carbon-vs-fossil.
34. *The Role of Critical Minerals in Clean Energy Transitions*, International Energy Agency, 2022, https://www.iea.org/reports/the-role-of-critical-minerals-in-clean-energy-transitions.
35. Seaver Wang et al., *Updated Mining Footprints and Raw Material Needs for Clean Energy*, Breakthrough Institute, 2024, https://thebreakthrough.org/issues/energy/updated-mining-footprints-and-raw-material-needs-for-clean-energy.
36. "Uranium Mining Overview," World Nuclear Association, May 16, 2024, https://world-nuclear.org/information-library/nuclear-fuel-cycle/mining-of-uranium/uranium-mining-overview.aspx.
37. "Backgrounder on Tritium, Radiation Protection Limits, and Drinking Water Standards," U.S. Nuclear Regulatory Commission, October 8, 2022, https://www.nrc.gov/reading-rm/doc-collections/fact-sheets/tritium-radiation-fs.html.
38. "Radionuclide Basics: Tritium," U.S. Environmental Protection Agency, updated June 4, 2024, https://www.epa.gov/radiation/radionuclide-basics-tritium.
39. "Radionuclide Basics: Tritium."
40. Mari Yamaguchi, "IAEA Officials Say Fukushima's Ongoing Release of Treated Radioactive Wastewater Is Going Well," *PBS News Hour*, October 23, 2023, https://www.pbs.org/newshour/world/iaea-officials-say-fukushimas-ongoing-release-of-treated-radioactive-wastewater-is-going-well.
41. "Fukushima Daiichi ALPS Treated Water Discharge—FAQs," International Atomic Energy Agency, accessed October 2022, https://www.iaea.org/topics/response/fukushima-daiichi-nuclear-accident/fukushima-daiichi-alps-treated-water-discharge/faq.
42. Navin Singh Khadka, "The Science Behind the Fukushima Waste Water Release," *BBC News*, August 25, 2023, https://www.bbc.com/news/world-asia-66610977.
43. Nick Touran and Delvan Neville, "Tuna Are Safe to Eat After Fukushima," *What Is Nuclear?* (blog), April 25, 2015, https://whatisnuclear.com/fukushima-fish.html.
44. Randall Munroe, "Spent Fuel Pool," *XKCD* (blog), 2022, https://what-if.xkcd.com/29.
45. "Nuclear Power in France," World Nuclear Association, updated May 21, 2024, https://world-nuclear.org/information-library/country-profiles/countries-a-f/france.aspx.
46. "Basic Information About Mercury," U.S. Environmental Protection Agency, accessed October 2024, https://www.epa.gov/mercury/basic-information-about-mercury.
47. Catherine Clifford, "The Energy in Nuclear Waste Could Power the U.S. for 100 Years, but the Technology Was Never Commercialized," CNBC, June 2, 2022, https://www.cnbc.com/2022/06/02/nuclear-waste-us-could-power-the-us-for-100-years.html.
48. Maddie Stone, "Solar Panels Are Starting to Die. What Will We Do with the Megatons of Toxic Trash?," *Grist*, August 13, 2020, https://grist.org/energy/solar-panels-are-starting-to-die-what-will-we-do-with-the-megatons-of-toxic-trash.

CHAPTER 5: DEGROWTH

1. S. Mintz and S. McNeil, "Housework in Late 19th Century America," *Digital History*, 2021, https://www.digitalhistory.uh.edu/topic_display.cfm?tcid=93.
2. Michele W. Berger, "How the Appliance Boom Moved More Women into the Workforce," *Penn Today*, January 30, 2019, https://penntoday.upenn.edu/news/how-appliance-boom-moved-more-women-workforce.
3. Jeremy Greenwood, *Evolving Households: The Imprint of Technology on Life* (MIT Press, 2019).
4. Nick Touran, "A Primer on Energy, Greenhouse Gas, Intermittency, and Nuclear," *What Is Nuclear?* (blog), December 17, 2017, https://whatisnuclear.com/primer-on-energy.html.
5. Matt Phillips and Andrew Freedman, "Right's New Fight: Gas Stoves," *Axios*, January 12, 2023, https://www.axios.com/2023/01/12/gas-stoves-conservative-backlash.
6. Ronny Jackson (@RonnyJacksonTX), "I'll NEVER give up my gas stove. If the maniacs in the White House come for my stove, they can pry it from my cold dead hands. COME AND TAKE IT!!," Twitter (now X), January 10, 2023, https://x.com/RonnyJacksonTX/status/1612839703018934274.
7. W. Wayt Gibbs, "How Much Energy Will the World Need?," *Anthropocene Magazine*, July 2017, https://www.anthropocenemagazine.org/howmuchenergy.
8. "Primary Energy Consumption Worldwide in 2023, by Country," Statista, June 2024, https://www.statista.com/statistics/263455/primary-energy-consumption-of-selected-countries.
9. Emily Elert, "If Everyone Lived Like an American, How Many Earths Would We Need?," *Popular Science*, October 20, 2012, https://www.popsci.com/environment/article/2012-10/daily-infographic-if-everyone-lived-american-how-many-earths-would-we-need.
10. Kavya Balaraman, "'A Tripling of Electrical Output': Tesla's Musk Urges Power Sector to Anticipate Higher Demand," *Utility Dive*, July 26, 2023, https://www.utilitydive.com/news/elon-musk-power-demand-growth-pge/688979.
11. Delger Erdenesanaa, "AI Could Soon Need as Much Electricity as an Entire Country," *New York Times*, October 10, 2023, https://www.nytimes.com/2023/10/10/climate/ai-could-soon-need-as-much-electricity-as-an-entire-country.html.
12. Evan Halper, "Amid Explosive Demand, America Is Running Out of Power," *Washington Post*, March 7, 2024, https://www.washingtonpost.com/business/2024/03/07/ai-data-centers-power.
13. "Global Primary Energy Consumption by Source," Our World in Data, accessed October 4, 2024, https://ourworldindata.org/grapher/global-energy-substitution.
14. Carol Rasmussen, "A Climate Conundrum: Why Didn't Atmospheric CO2 Fall During the Pandemic?," *Caltech Magazine*, March 30, 2022, https://magazine.caltech.edu/post/atmospheric-co2-covid-pandemic.
15. Jessica Evans, "NASA Science Enables First-of-Its-Kind Detection of Reduced Human CO2 Emissions," *NASA Explore*, March 31, 2022, https://www.nasa.gov/earth-and-climate/nasa-science-enables-first-of-its-kind-detection-of-reduced-human-co2-emissions/.

CHAPTER 6: CLEAN ENERGY REVOLUTION

1. Bent Sørensen, "A New Energy Source," *Science* 189, no. 4199 (1975): 255, https://archive.org/details/sim_science_1975-07-25_189_4199/page/255.
2. Ken Silverstein, "Are Fossil Fuel Interests Bankrolling the Anti-Nuclear Energy Movement?," *Forbes*, July 13, 2016, https://www.forbes.com/sites/kensilverstein/2016/07/13/are-fossil-fuel-interests-bankrolling-the-anti-nuclear-energy-movement/?sh=358c77737453.
3. Alan Chodos, "April 25, 1954: Bell Labs Demonstrates the First Practical Silicon Solar Cell," *APS News*, April 1, 2009, https://www.aps.org/publications/apsnews/200904/physicshistory.cfm.
4. Lucía Fernández, "Share of Electricity Generation from Solar Energy Worldwide from 2010 to 2023," Statista, June 24, 2024, https://www.statista.com/statistics/1302055/global-solar-energy-share-electricity-mix.
5. "What Is U.S. Electricity Generation by Energy Source?," U.S. Energy Information Administration, February 2024, https://www.eia.gov/tools/faqs/faq.php?id=427&t=3.
6. Ember and Energy Institute, "Share of Electricity Generated by Wind Power," Our World in Data, 2024, https://ourworldindata.org/grapher/share-electricity-wind?tab=chart.
7. "What Is U.S. Electricity Generation by Energy Source?"
8. Ember and Energy Institute, "Share of Electricity Generated by Hydropower," https://ourworldindata.org/grapher/share-electricity-hydro.
9. Christine Shearer et al., "Geothermal Power," Project Drawdown, accessed October 4, 2024, https://drawdown.org/solutions/geothermal-power.
10. Anne Brock, "How Do Clouds Affect Solar Panels?," *Solar Alliance* (blog), updated January 23, 2023, https://www.solaralliance.com/how-do-clouds-affect-solar-panels.
11. Matt Oliver, "'Dunkelflaute' Sends Wind Power Generation Plummeting in UK and Germany," *Telegraph*, November 2024, https://www.telegraph.co.uk/business/2024/11/05/dunkelflaute-cut-wind-power-generation-germany-uk.
12. Gavin Maguire, "Reduced Wind Generation Puts Texas Power System to the Test," *EnergyNow Media*, June 21, 2023, https://energynow.com/2023/06/column-reduced-wind-generation-puts-texas-power-system-to-the.test.
13. Wesley Cole and Akash Karmakar, *Cost Projections for Utility-Scale Battery Storage: 2023 Update*, National Renewable Energy Laboratory, 2023, https://www.nrel.gov/docs/fy23osti/85332.pdf; Paul Denholm, Wesley Cole, and Nate Blair, "Moving Beyond 4-Hour Li-Ion Batteries: Challenges and Opportunities for Long(er)-Duration Energy Storage," National Renewable Energy Laboratory, 2023, https://www.nrel.gov/docs/fy23osti/85878.pdf.
14. Nestor A. Sepulveda et al., "The Design Space for Long-Duration Energy Storage in Decarbonized Power Systems," *Nature Energy* 6 (2021): 506–16, https://www.nature.com/articles/s41560-021-00796-8.
15. Mark Z. Jacobson, "Zero Air Pollution and Zero Carbon from All Energy Without Blackouts at Low Cost in Texas," Stanford University, December 7, 2021, https://web.stanford.edu/group/efmh/jacobson/Articles/I/21-USStates-PDFs/21-WWS-Texas.pdf.

16. *Lithium-Ion Battery Manufacturing Capacity, 2022–2030*, International Energy Agency, 2023, https://www.iea.org/data-and-statistics/charts/lithium-ion-battery-manufacturing-capacity-2022-2030.
17. "Country Profile: Sweden," World Nuclear Association, updated September 2023, https://world-nuclear.org/information-library/country-profiles/countries-o-s/sweden.aspx.
18. Emily Chung, "How Sweden Electrified Its Home Heating—and What Canada Could Learn," *CBC News*, April 12, 2023, https://www.cbc.ca/news/science/sweden-heat-pumps-1.6806799.
19. Ryan Cooper, "Why America Should Double Down on Hydropower," *The Week*, January 22, 2019, https://theweek.com/articles/818273/why-america-should-double-down-hydropower; "U.S. Hydropower Market Report," U.S. Department of Energy, 2023, https://www.energy.gov/sites/default/files/2023-09/U.S.%20Hydropower%20Market%20Report%202023%20Edition.pdf.

CHAPTER 7: CHANGING TIDES

1. Spencer R. Weart, *The Rise of Nuclear Fear* (Harvard University Press, 2019).
2. Roger M. Macklis, "The Great Radium Scandal," *Scientific American* 296, no. 2 (August 1993).
3. David Burnham, "Nuclear Experts Debate 'The China Syndrome,'" *New York Times*, March 18, 1979, https://www.nytimes.com/1979/03/18/archives/nuclear-experts-debate-the-china-syndrome-but-does-it-satisfy-the.html.
4. Diana Kruzman, "Wood-Burning Stoves Raise New Health Concerns," *Undark*, March 2, 2022, https://undark.org/2022/03/02/wood-burning-stoves-raise-new-health-concerns.
5. Liz Ruskin and Emily Holden, "Natural but Deadly: Huge Gaps in U.S. Rules for Wood-Stove Smoke Exposed," *Guardian*, March 16, 2021, https://www.theguardian.com/environment/2021/mar/16/wood-smoke-alaska-state-regulators-air-quality.
6. Sonal Patel, "Global Report Warns of Looming Skills Shortages in Power, Nuclear, Renewables Sectors," *Power Magazine*, January 24, 2019, https://www.powermag.com/global-report-warns-of-looming-skills-shortages-in-power-nuclear-renewables-sectors; Global Energy Talent Index, 2019, https://cdn2.hubspot.net/hubfs/3277184/GETI/GETI%202019/Global%20Energy%20Talent%20Index%20-%20GETI%202019.pdf.
7. Benjamin Storrow, "3 States with Shuttered Nuclear Plants See Emissions Rise," *E&E News by Politico*, February 17, 2022, https://www.eenews.net/articles/3-states-with-shuttered-nuclear-plants-see-emissions-rise.
8. "*The Simpsons* May Affect Views of Nuclear Plants, Professor Says," *CBC News*, February 24, 2022, https://www.cbc.ca/news/entertainment/the-simpsons-may-affect-view-of-nuclear-plants-prof-1.807902.
9. Bryan Walsh, "Exclusive: How the Sierra Club Took Millions from the Natural Gas Industry—and Why They Stopped," *TIME*, February 2, 2012, https://science.time.com/2012/02/02/exclusive-how-the-sierra-club-took-millions-from-the-natural-gas-industry-and-why-they-stopped.

10. Brock Cooper, "Is Nuclear Energy Worth the Risk?," *News Tribune*, July 20, 2007, https://web.archive.org/web/20110714174817/http:/www.newstrib.com/featured-series/energy-series/Articles/A_7-20-2007_1_4.pdf.
11. "Greenpeace and Utilities Launch Suit Against Hinkley Nuclear Plant," *Guardian*, July 2, 2015, https://www.theguardian.com/environment/2015/jul/02/greenpeace-utilities-launch-suit-against-hinkley-nuclear-plant.
12. Lachlan Markay, "Inside the Gas Industry's Plan to Sink Nuclear Power," *Daily Beast*, July 1, 2023, https://www.thedailybeast.com/inside-the-gas-industrys-plan-to-sink-nuclear-power?ref=scroll.

CHAPTER 8: THREE DECADES OF BAD ENERGY POLICY

1. Andreas Wilkens, "Energy Industry: Energy Transition Will Cost 1.2 Trillion Euros by 2035," *Heise Online*, April 30, 2024, https://www.heise.de/en/news/Energiewirtschaft-Energiewende-kostet-bis-2035-1-2-Billionen-Euro-9703870.html.
2. *Germany: Falling Behind in the Electricity Transition*, Ember, 2022, https://ember-climate.org/app/uploads/2022/02/NECP-Factsheet-Germany.pdf.
3. Zia Weise, "Why Germany Won't Give Up on Giving Up Nuclear," *Politico*, April 28, 2022, https://www.politico.eu/article/politics-behind-germany-refusal-reconsider-nuclear-phaseout.
4. Sarah White et al., "The Nuclear Dispute Driving a Wedge Between France and Germany," *Financial Times*, October 14, 2023, https://www.ft.com/content/b1dbd7b4-d8b9-45eb-bd18-4976f7c9af5e.
5. "Germany to Postpone Nuclear Plant Closures as Russia's War in Ukraine Fuels Energy Crisis Fears," *CBC News*, October 17, 2022, https://www.cbc.ca/news/world/germany-nuclear-plants-april-1.6619144.
6. Katrin Bennhold, "The Former Chancellor Who Became Putin's Man in Germany," *New York Times*, April 23, 2022, https://www.nytimes.com/2022/04/23/world/europe/schroder-germany-russia-gas-ukraine-war-energy.html.
7. Katrin Bennhold, "The End of the (Pipe)line? Germany Scrambles to Wean Itself Off Russian Gas," *New York Times*, updated April 21, 2022, https://www.nytimes.com/2022/04/06/world/europe/germany-gas-russia-ukraine.html.
8. Jillian Weinberger, Amy Drozdowska, and Byrd Pinkerton, "How Germany Helped Make Renewable Energy Cheap for the Rest of the World," *Vox*, January 22, 2020, https://www.vox.com/2020/1/22/21028914/germany-green-new-deal-solar-power.
9. Julian Wettengel, "IEA Head Says German Nuclear Exit 'Historic Mistake,' Criticises Bet on Hydrogen in Short Term," *Clean Energy Wire*, January 23, 2024, https://www.cleanenergywire.org/news/iea-head-says-german-nuclear-exit-historic-mistake-criticises-bet-hydrogen-short-term.
10. Frédéric Simon, "Seven Countries Reject Nuclear-Derived Hydrogen from EU Renewables Law," *Euractiv*, March 20, 2023, https://www.euractiv.com/section/energy-environment/news/seven-countries-reject-nuclear-derived-hydrogen-from-eu-renewables-law.
11. "France Tops Europe's Net Power Export Chart," *Power Engineering International*, February 6, 2024, https://www.powerengineeringint.com/world-regions/europe/france-tops-europes-net-power-export-chart/.

12. Peter Fairley, "Renewables Up, Nuclear Down in French Energy Plan," *IEEE Spectrum*, June 24, 2014, https://spectrum.ieee.org/renewables-up-nuclear-down-in-french-energy-plan.
13. Paul Day, "Vogtle's Troubles Bring U.S. Nuclear Challenge into Focus," Reuters, August 24, 2023, https://www.reuters.com/business/energy/vogtles-troubles-bring-us-nuclear-challenge-into-focus-2023-08-24.

CHAPTER 9: VIBE SHIFT

1. Dan Walters, "The Brown Link to Indonesian Firm," *Lodi News-Sentinel*, October 17, 1990, https://news.google.com/newspapers?nid=2245&dat=19901017&id=HhYzAAAAIBAJ&sjid=MTIHAAAAIBAJ&pg=4793,5589478.
2. Rebecca Tuhus-Dubrow, "The Activists Who Embrace Nuclear Power," *New Yorker*, February 19, 2021, https://www.newyorker.com/tech/annals-of-technology/the-activists-who-embrace-nuclear-power.
3. Carolyn Porco, "Love the Earth—Nuclear Power to the People," *Coming of Age in the Solar System* (blog), March 18, 2024, https://carolynporco.substack.com/p/nuclear-power-to-the-people.
4. Carolyn Porco, "Extensive Scientific Scrutiny Tells Us Cassini Is Safe," *Arizona Daily Star*, October 7, 1997, http://carolynporco.com/in-the-news/popular-writings/arizona-daily-star-cassini-is-safe-carolyn-porco.pdf.
5. "Public Support for Nuclear Energy Is Highest Among Plant Neighbors," American Nuclear Society, August 16, 2022, https://www.ans.org/news/article-4228/public-support-for-nuclear-energy-is-highest-among-plant-neighbors.
6. Ann S. Bisconti, "Reverse NIMBY: Nuclear Power Plant Neighbors Say 'Yes,'" Bisconti Research Inc., June 2022, https://www.bisconti.com/blog/9th-national-survey-of-nuclear-power-plant-neighbors.
7. Megan Brenan, "Americans' Support for Nuclear Energy Is at Highest in a Decade," Gallup, April 6, 2023, https://news.gallup.com/poll/474650/americans-support-nuclear-energy-highest-decade.aspx.
8. Robert Bryce, "The Anti-Nuclear Industry Is a $2.3B-Per-Year Racket," *Robert Bryce* (blog), August 24, 2023, https://robertbryce.substack.com/p/the-anti-nuclear-industry-is-a-23b.
9. "At COP28, Countries Launch Declaration to Triple Nuclear Energy Capacity by 2050, Recognizing the Key Role of Nuclear Energy in Reaching Net Zero" U.S. Department of Energy, December 1, 2023, https://www.energy.gov/articles/cop28-countries-launch-declaration-triple-nuclear-energy-capacity-2050-recognizing-key.
10. "How 'World View' Affects Public Perception of Nuclear Power," *World Nuclear News*, October 18, 2019, https://www.world-nuclear-news.org/Articles/How-world-view-affects-public-perception-of-nuclea.
11. Josh Siegel, "GOP Embraces Liberal Environmentalism Skeptic with Radical Plan for Nuclear Energy," *Washington Examiner*, July 28, 2020, https://www.washingtonexaminer.com/news/2372675/gop-embraces-liberal-environmentalism-skeptic-with-radical-plan-for-nuclear-energy.
12. Jessica Lovering, speech given at International Conference on Climate Change and the Role of Nuclear Power, 2019, International Atomic Energy Agency, Vienna.

13. Anne-Sophie Hacquin et al., "Disgust Sensitivity and Public Opinion on Nuclear Energy," *Journal of Environmental Psychology* 80 (April 2022), https://www.sciencedirect.com/science/article/abs/pii/S0272494421002024.
14. James Titcomb, "Silicon Valley Finds Its Silver Bullet in a Desperate Race for Energy," *Telegraph*, May 5, 2024, https://www.telegraph.co.uk/business/2024/05/05/ai-boom-nuclear-power-electricity-demand/.
15. Paul Hockenos, "Poland Is Dreaming of Becoming a Nuclear Power," *Foreign Policy*, October 17, 2023, https://foreignpolicy.com/2023/10/17/poland-government-civic-platform-pis-tusk-nuclear-energy.
16. "U.S. Announces New Support for Ghana's Civil Nuclear Energy Program," U.S. Embassy in Ghana, September 13, 2023, https://gh.usembassy.gov/u-s-announces-new-support-for-ghanas-civil-nuclear-energy-program-under-the-first-capacity-building-program.
17. Paul Day, "Vogtle's Troubles Bring U.S. Nuclear Challenge into Focus," Reuters, August 24, 2023, https://www.reuters.com/business/energy/vogtles-troubles-bring-us-nuclear-challenge-into-focus-2023-08-24.
18. Jeff Amy, "The First U.S. Nuclear Reactor Built from Scratch in Decades Enters Commercial Operation in Georgia," Associated Press, July 31, 2023, https://apnews.com/article/georgia-power-nuclear-reactor-vogtle-9555e3f9169f2d58161056feaa81a425.
19. Hannah Ritchie, "Solar Panel Prices Have Fallen by Around 20% Every Time Global Capacity Doubled," Our World in Data, June 12, 2024, https://ourworldindata.org/data-insights/solar-panel-prices-have-fallen-by-around-20-every-time-global-capacity-doubled.
20. Spencer Kimball, "Oracle Is Designing a Data Center That Would Be Powered by Three Small Nuclear Reactors," *CNBC News*, September 10, 2024, https://www.cnbc.com/2024/09/10/oracle-is-designing-a-data-center-that-would-be-powered-by-three-small-nuclear-reactors.html.
21. "Powering Intelligence: Analyzing Artificial Intelligence and Data Center Energy Consumption," Electric Power Research Institute, 2024, https://www.epri.com/research/products/3002028905.

ILLUSTRATION CREDITS

CHAPTER 2: HUMANS AND ENERGY
p. 23: "Global Primary Energy Consumption by Source," Our World in Data, 2024, https://ourworldindata.org/grapher/global-energy-consumption-source.
p. 26: "U.S. Energy Facts Explained," U.S. Energy Information Administration, updated September 2023, https://www.eia.gov/energyexplained/us-energy-facts.
p. 27: Hannah Ritchie and Pablo Rosado, "Electricity Mix," Our World in Data, updated January 2024, https://ourworldindata.org/electricity-mix.

CHAPTER 3: FOSSIL FUELS' FINAL BOSS
p. 31: "Coal Explained: Coal and the Environment," U.S. Energy Information Administration, April 17, 2024, https://www.eia.gov/energyexplained/coal/coal-and-the-environment.php.
p. 32: "Nuclear Fuel and Its Fabrication," World Nuclear Association, October 13, 2021, https://world-nuclear.org/information-library/nuclear-fuel-cycle/conversion-enrichment-and-fabrication/fuel-fabrication.
p. 36: "Pressurized Water Reactors," U.S. Nuclear Regulatory Commission, updated February 9, 2023, https://www.nrc.gov/reactors/power/pwrs.html.
p. 39: Hannah Ritchie and Pablo Rosado, "Electricity Mix," Our World in Data, updated January 2024, https://ourworldindata.org/electricity-mix.
p. 40: "Electric Power Monthly," U.S. Energy Information Administration, accessed April 2024, https://www.eia.gov/electricity/monthly/.
p. 42: "3 Reasons Why Nuclear Is Clean and Sustainable," U.S. Department of Energy, updated June 2022, https://www.energy.gov/ne/articles/3-reasons-why-nuclear-clean-and-sustainable.
p. 43: Cody Good, "How Much Land Does Electricity Use?," *Energy Minute*, August 18, 2023, https://energyminute.ca/infographics/how-much-land-does-electricity-use/.
p. 44: Justin Aborn et al., "An Assessment of the Diablo Canyon Nuclear Plant for Zero Carbon Electricity, Desalination and Hydrogen Production," MIT Center for Energy & Environmental Policy Research, 2021, https://drive.google.com/file/d

/1RcWmKwqgzvIgIlhOBB2s5cA6ajuVJJzt/view; "Global Methane Tracker 2024," International Energy Agency, 2024, https://www.iea.org/reports/global-methane-tracker-2024/key-findings.

p. 46, top: "Ontario's System-Wide Electricity Supply Mix: 2003 Data," Ontario Energy Board, October 26, 2005, https://www.oeb.ca/documents/electricity_mix_261005.pdf.

p. 46, bottom: "Ontario's System-Wide Electricity Supply Mix: 2023 Data," Ontario Energy Board, May 22, 2024, https://www.oeb.ca/sites/default/files/2023-supply-mix-data-update.pdf.

p. 47: Hannah Ritchie and Pablo Rosado, "Electricity Mix."

p. 49: "Global Methane Tracker 2024."

p. 51: Nichola Groom, "Nuclear, Coal, Oil Jobs Pay More Than Those in Wind, Solar—Report," Reuters, April 16, 2021, https://www.reuters.com/article/idUSKBN2BT2OO.

p. 56: "Carbon Neutrality in the UNECE Region: Integrated Life-cycle Assessment of Electricity Sources," United Nations Economic Commission for Europe, 2022, https://unece.org/sites/default/files/2022-08/LCA_0708_correction.pdf.

CHAPTER 4: DISPELLING MYTHS

p. 64: "Deaths per TWh Energy Production," Our World in Data, accessed 2024, https://ourworldindata.org/grapher/death-rates-from-energy-production-per-twh.

p. 77: "Kiggavik Project Environmental Impact Statement," Areva Resources Canada, Inc., 2011, https://makitanunavut.wordpress.com/wp-content/uploads/2012/05/deis-volume-2-project-description.pdf.

p. 79: Hannah Ritchie, "Mining Quantities for Low-Carbon Energy Is Hundreds to Thousands of Times Lower Than Mining for Fossil Fuels," *Sustainability by Numbers* (blog), January 18, 2023, https://www.sustainabilitybynumbers.com/p/mining-low-carbon-vs-fossil.

p. 80: Seaver Wang et al., "Updated Mining Footprints and Raw Material Needs for Clean Energy," Breakthrough Institute, 2024, https://thebreakthrough.org/issues/energy/updated-mining-footprints-and-raw-material-needs-for-clean-energy.

p. 84: "Tritium in Exit Signs," Environmental Protection Agency, November 16, 2023, https://www.epa.gov/radtown/tritium-exit-signs.

p. 88: Mahendra Jivanlal Shah et al., "Tip-Over Analysis of the HI-STORM Dry Storage Cask System," January 2003, https://www.researchgate.net/publication/237278241_Tip-Over_Analysis_of_the_HI-STORM_Dry_Storage_Cask_System.

p. 91: "Final Disposal: Disposal Canister," Posiva, accessed December 2024, https://www.posiva.fi/en/index/finaldisposal/releasebarriers/disposalcanister.html.

p. 94: Solomon Goldstein-Rose, *The 100% Solution: A Plan for Solving Climate Change* (Melville House, 2020).

CHAPTER 5: DEGROWTH

p. 99: Nick Touran, "A Primer on Energy, Greenhouse Gas, Intermittency, and Nuclear," *What Is Nuclear?* (blog), December 17, 2017, https://whatisnuclear.com/primer-on-energy.html.

p. 103: Srini Bangalore et al., "Investing in the Rising Data Center Economy," McKinsey and Company, January 2023, https://www.mckinsey.com/industries/tech

nology-media-and-telecommunications/our-insights/investing-in-the-rising
-data-center-economy.
p. 105: Hannah Ritchie, *Not the End of the World* (Little, Brown Spark, 2024).

CHAPTER 6: CLEAN ENERGY REVOLUTION
p. 116: Hannah Ritchie and Pablo Rosado, "Electricity Mix," Our World in Data, updated January 2024, https://ourworldindata.org/electricity-mix.
p. 121: Mark Z. Jacobson, "Zero Air Pollution and Zero Carbon from All Energy Without Blackouts at Low Cost in Texas," Stanford University, December 7, 2021, https://web.stanford.edu/group/efmh/jacobson/Articles/I/21-USStates-PDFs/21-WWS-Texas.pdf.
p. 123: Hannah Ritchie and Pablo Rosado, "Electricity Mix."
p. 124: Hannah Ritchie and Pablo Rosado, "Electricity Mix."
p. 125: Hannah Ritchie and Pablo Rosado, "Electricity Mix."
p. 126: Hannah Ritchie and Pablo Rosado, "Electricity Mix."
p. 127: Hannah Ritchie and Pablo Rosado, "Electricity Mix."
p. 128: "Ontario's System-Wide Electricity Supply Mix: 2023 Data," Ontario Energy Board, May 22, 2024, https://www.oeb.ca/sites/default/files/2023-supply-mix-data-update.pdf.

CHAPTER 7: CHANGING TIDES
p. 140: Tom Parnell from Scottish Borders, Scotland, CC BY-SA 2.0. https://creativecommons.org/licenses/by-sa/2.0, via Wikimedia Commons.
p. 149: Oliver Milman, "A Nuclear Plant's Closure Was Hailed as a Green Win. Then Emissions Went Up," *Guardian*, March 20, 2024, https://www.theguardian.com/environment/2024/mar/20/nuclear-plant-closure-carbon-emissions-new-york.
p. 153: T. Maher CFMEU Mining Division, "Protecting Mining Jobs and Communities," advertisement, *Courier Mail*, November 19, 2007, https://thoriumenergyalliance.com/wp-content/uploads/2020/02/anti-nuclear_coal_ad_md.jpg.

CHAPTER 8: THREE DECADES OF BAD ENERGY POLICY
p. 159: "Carbon Intensity of Electricity Generation," Our World in Data, 2024, https://ourworldindata.org/grapher/carbon-intensity-electricity.
p. 162: Hannah Ritchie, "Data on the German Retreat from Nuclear Energy Tell a Cautionary Tale," *Washington Post*, May 10, 2023, https://www.washingtonpost.com/opinions/2023/05/10/germany-end-nuclear-cost-climate-health/.
p. 164: Hannah Ritchie and Max Roser, "France: CO2 Country Profile," Our World in Data, 2020, https://ourworldindata.org/co2-and-greenhouse-gas-emissions.

CHAPTER 9: VIBE SHIFT
p. 181: Megan Brenan, "Nuclear Energy Support Near Record High in U.S.," Gallup, April 9, 2025, https://news.gallup.com/poll/659180/nuclear-energy-support-near-record-high.aspx.
p. 182: Richard Ollington, "Public Attitudes Toward Clean Energy—Nuclear," Radiant Energy Group, 2023, https://www.radiantenergygroup.com/reports/public-attitudes-toward-clean-energy-2023-nuclear.

Illustration Credits

p. 184: "Country Analysis Brief: Japan," U.S. Energy Information Administration, updated July 7, 2023, https://www.eia.gov/international/content/analysis/countries_long/japan/japan.pdf.

p. 187: Richard Ollington, "Public Attitudes Toward Clean Energy—Nuclear."

p. 192: Alex de Vries, "The Growing Energy Footprint of Artificial Intelligence," *Joule* 7, no. 10 (October 18, 2023), https://www.sciencedirect.com/science/article/pii/S2542435123003653.

INDEX

Note: Italicized page numbers indicate material in photographs or illustrations.

Aalo Atomics, 196
Abe, Shinzo, 183
agriculture, xiv–xv
air-conditioning, 100, 128, 175
air quality, 6, 11, 24–25, 92. See also carbon emissions
air travel, 70, 71
Aktau, Kazakhstan, 57
alpha emitters, 73
Amarakaeri people, 77
Amazon (company), 196
Amazon (rainforest), xiii–xiv, 207
Amazon Web Services, 191
Amchitka Island, 145
American Petroleum Institute, 155
Amnesty International, 78
Anderson, Robert Orville, 109
antihuman rhetoric, xv–xvi, 202
antinuclear activism, xx, 143–51, 151–56, 157, 165, 200–201
Arizona, 57
arms race, 9–10
artificial intelligence (AI)
    energy demands of, 103–4, 104n, 189, 191–93, 198
    and future prospects for nuclear, 201
    and potential of nuclear power, 7
    power requirements, 192
    See also ISODOPE

Atlantic Richfield, 109
atomic bombs. See nuclear weapons
Atomic Energy Commission, 166
Atomium, 139, 140
atoms, structure of, 29, 29n, 30, 33, 35, 213
Atoms for Peace, 138–39, 141
Attenborough, David, xv
Aurora reactor design, 196
Australia, xiii, 153, 182, 207

background radiation, 70
bananas, 69n, 70
Banqiao Dam, 61, 63, 67–68
barium, 4–5
batteries
    and definitions of energy, 111, 215
    and energy diet concept, 119–21, 121, 128
    and mined resources, 77–78, 80
    radioisotope thermoelectric generators, 177, 177n
    and resource requirements of energy sources, 202–3
The Beginning of Infinity (Deutsch), xvi, 13
Belgium, 139, 140
Biden, Joe, 189
big bang, 20
Bikini Atoll, 142
biodiversity, xiv, 25

bioenergy
   biofuels, *23*
   biomass energy, 22, *23*, *64*, 213
   and France's energy sources, *47*
   and global energy sources, *27*, *124*
   and Japan's energy sources, *184*
   and Ontario's energy sources, *46*, *128*
   and sources of methane emissions, *49*
   and Switzerland's energy sources, *127*
   and U.S.'s energy sources, *125*
Birol, Fatih, 163
birth defects, 72–74
Brazil
   and author's background, xiii–xv, xvii, 98–100, 176
   and energy poverty, 11–12
   and public perception of nuclear power, 182, *182*
Brazil nuts, 69
Breakthrough Institute, 78
British Petroleum, 101
Brower, David, 109, 172–73
Brown, Jerry, 173–74
Browne, Jackson, 147, 173
Byers, Eben, 135

California
   and author's background, xiii
   and capacity factor of energy sources, 41
   and degrowth ideology, 97
   and nuclear-powered desalination, 57
   and resource requirements of energy sources, 43
   and shifting sentiment on nuclear, 172–81
   and solar energy, 161
   and tech industry's demand for power, 192
   wildfires, 207
   See also Diablo Canyon
Cal Poly State University, 178
cancers, 74, 76
capacity factor, 39–40, *40*, 161, 213
capitalism, 98–99
carbon emissions
   advantages of nuclear power, 47
   and author's environmental awakening, xiv
   and carbon footprint, 214
   and climate change, 25
   and closure of nuclear facilities, 149
   and demand for electricity, 105
   European countries compared, *159*
   fossil fuels contrasted with nuclear, 178
   and Germany's *Energiewende* program, 158n
   and greenhouse gases, 216
   impact of Messmer Plan, *164*
   and life cycle impact of energy sources, 55
   and mined resources, 79
   New York, *149*
   relation to GDP, *105*
   and waste levels of power sources, 92–93
carbon footprint, xv, 55, 101, 214
Carbon Free California, 176
carbon tax, 122
cars, nuclear, 139, *139*
Carson, Rachel, 143
Cassini spacecraft, 177
Centers for Disease Control and Prevention, 76
chain reaction, 32, 35, 53, 65, 86n, 137, 214. *See also* fission
charcoal, 24, 24n
ChatGPT, 192, *192*
Chernobyl (HBO series), 72–73
Chernobyl nuclear accident
   and antinuclear sentiment, 159–60, 168
   immediate and long-term effects, 72–74, 95
   impact on Italy's energy policy, 190
   lack of containment structure, 36n
   myths and misconceptions, 61–68
   and public perception of nuclear power, xix, 59, 68, 142, 148
   and support for nuclear in Ukraine, 183
Chicago Pile-1, 6
child labor, 78
Chile, 77–78
China, 168, *182*, 188–90, 196
The China Syndrome (film), 147, 157
chronic obstructive pulmonary disease, 24
Chu, Steven, 179
Clean Air and Clean Water Acts, 144
Clean Air Task Force, 43, 176, 211

Index

"clean coal" technology, 153
clean energy
   and antinuclear sentiment, 163
   and capacity factor of energy sources, 41
   and definition of renewable energy, 217
   and energy diet concept, 117, 120, 130
   and France's energy sources, 46
   and Georgia's energy sources, 190
   geothermal power as, 115–16
   and global energy sources, 29, *116*
   hydropower as, 114–15
   and mined resources, 78–80
   and origins of antinuclear movement, 154
   and potential of nuclear power, 7, 15, 199–200
   and promotion of nuclear energy, 204–5, 208
   and public perception of nuclear power, 60, 175, 178–79
   "renewable" energy described, 108–12
   repurposing fossil fuel plants, 49
   and resource requirements of energy sources, 42–44, 203
   and social benefits of transition to nuclear, 51
   solar power as, 112–13
   and U.S.'s energy sources, *38*, 38–39
   wind power as, 113–14
   See also geothermal power; hydropower; solar power; wind power
climate change, 14–15, 25. *See also* global warming
CO_2. *See* carbon emissions
coal and coal power
   and antinuclear sentiment, 162, 167
   and capacity factor of energy sources, *40*
   and causes of Chernobyl disaster, 65
   and closure of nuclear facilities, 149
   contrasted with nuclear, *31*
   deaths associated with power sources, *64*
   definition of fossil fuels, 215
   and electricity production/consumption, *26*
   and energy density of power sources, *42*
   and energy diet concept, 118, 129
   and France's energy sources, *47*
   and gender gap in support for nuclear, 188
   and global energy sources, *23*, 27, *27*, *124*
   historical background of power sources, 97
   and history of energy sources, 23–26
   and Japan's energy sources, *184*
   and life cycle impact of energy sources, 55, *56*
   and mercury waste, 89
   mining, 77–78, *79*, *80*, 136, 153
   and Ontario's energy sources, 45, *46*
   and origins of antinuclear movement, 152
   and potential of nuclear power, 7, 11
   radiation associated with, 72, 72n
   and relative dangers of energy sources, 63
   repurposing fossil fuel plants, 49–50, 58
   and resource requirements of energy sources, 41–42, *43*
   and response to Fukushima disaster, 68
   and small modular reactor development, 194–95
   and social benefits of transition to nuclear, 50–51
   and sources of methane emissions, 49
   and U.S.'s energy sources, *125*
   and waste levels of power sources, 92–93, *94*
cobalt, 78
Cold War
   and historical background of nuclear, 133
   impact on nuclear's potential, 9, 200
   and potential of nuclear power, 11
   and public perception of nuclear power, xix, 141
Columbia University, 142
Constellation, 191, 191n
containment structures, 36, *36*, 36n, 214
cooling systems, 36–38, *38*, 57, 66n
cosmic radiation, 20, 63, 69, 218
counterculture movement, 144
COVID-19 pandemic, xv, 104–5, 167
CT scans, *70*
Cuban Missile Crisis, 142–43
Curie, Marie, xi
Cuyahoga River, 144

Index

The Daily Beast, 155
data centers, 103, 103–4, 191–93
Dawkins, Richard, xvii
deaths associated with power sources
   and Chernobyl nuclear accident, 61–63
   and Fukushima disaster, 67–68
   and historical background of nuclear, 136–37
   and mining dangers, 77
   and potential of nuclear power, 6, 11
   and public perception of nuclear power, 148
   and radiation exposure, 71
   and Three Mile Island, 147
decarbonization, 141, 158n, 159, 163
decommissioning, 48, 50, 56
deep boreholes, 91–92
deforestation, xiii–xiv, 24
degrowth, 96, 97–102, 102–6, 214
demand for electricity, 102–6, 175
Democratic Republic of the Congo, 78
Denmark, 73
density of power sources
   advantages of nuclear power, 48
   and energy diet concept, 119
   and mined resources, 78
   and potential of nuclear power, 5
   and repurposing of fossil fuel infrastructure, 50
   and resource requirements of energy sources, 41–42
desalination, 44, 57
deuterium, 82
Deutsch, David, xvi, 13
developing nations, 100–102
Diablo Canyon
   and author's pro-nuclear efforts, xxi
   and nuclear-powered desalination, 57
   and resource requirements of energy sources, 43, *44*
   and shifting sentiment on nuclear, 172–81, *180*
Disney, Walt, 140
Don't Make a Wave Committee, 145–46
donut shop (analogy), 39–40
dry cask storage, *85*, 87–90, *88*, 95, *151*, 203–4, 214

Earth Day, 144
earthquakes, 66–67. See also Fukushima (Daichi)
Eaton Fire, 207
economic benefits of nuclear, 7, 14, 50–51
Einstein, Albert, 9
Eisenhower, Dwight, 138
electricity and electrification
   demand for, 102–6
   and economic benefits of nuclear, 7–8
   labor-saving benefits of, 97–102, *98*
   and quality of life, *99*
   sources and uses of, 25–27, *26*
   See also specific energy sources
Electric Power Research Institute, 192
electrons, 29, 29n, *30*, 113n, 213, 215
Ellison, Larry, 191
Energiewende program, 158–59, 158n
energy (definition), 19–22, 215
energy diet, 117–29, *123–28*
energy timeline, 22–27
Enewetak Atoll, 142
Environmental Progress, 173–74
Environmental Protection Agency, 76n, 82, 144
European Commission, 154
European Court of Justice, 154
evolution, xvii–xviii, 21
exclusion zone (Chernobyl), 66
Exxon, 151

Fermi, Enrico, 3
fertilizers, 44
financial crisis of 2009, 104
financing for nuclear projects, 189–90
Finland, 91, *105*
fire, 21–22, 54, 54n, 207
fission
   early research on, 3–5
   and historical background of nuclear power, 134
   neutron's role in, 216
   and nuclear power generation, 36
   and origins of antinuclear movement, 155
   and potential of nuclear power, 5–8
   process described, 4–5, 29–30, *30*, 35, 214, 215

Index

and public perception of nuclear power, 140–41
and secrecy of nuclear research, 137–38
and spent nuclear fuel, 86n
and tritium production, 81
and uranium mining, 75
Florida Power and Light, 169
Fonda, Jane, 147, 165
food chain, 20–21
fossil fuels
  alternatives to, 29
  and antinuclear sentiment, 161, 167, 170
  and author's pronuclear efforts, xx, xxii
  carbon emissions contrasted with nuclear, 178
  and carbon footprint concept, 101
  and construction of nuclear plants, 48
  defined, 215
  and definition of renewable energy, 111
  and degrowth ideology, 98, 103
  and energy diet concept, 117–18, 122, 129, 130
  and France's energy sources, *47*, 164–65
  and gender gap in support for nuclear, 186
  and Germany's *Energiewende* program, 158–59, 158n
  and Germany's energy sources, 158–63, *162*
  and global energy sources, *23*, *27*
  and greenhouse gases, 216
  and history of energy sources, 23–27, 28
  impact on nuclear's potential, 200
  Japan's energy policy after Fukushima, 184
  and mined resources, 77–79
  and necessity of energy transition, 53
  and Ontario's energy sources, 45–46, *46*
  and origins of antinuclear movement, 108–9, 146, 151–56, 157
  and Poland's energy sources, 190
  and potential of nuclear power, 6–7, 11–12, 15
  and promotion of nuclear energy, 205–6, 208
  and relative dangers of energy sources, 63
  and resource requirements of energy sources, 41, 44, 203
  and response to Fukushima disaster, 68
  and shifting sentiment on nuclear, 171–72
  and submarines, 194
  and tech industry's demand for power, 193
  and U.S.'s energy sources, 125
  and waste levels of power sources, 92–93
  See also specific fossil fuel types
fracking, 195
France
  and antinuclear sentiment, 160–63
  benefits of nuclear power in, 163–65, 170
  and capacity factor of energy sources, 40
  carbon emission of European countries, *159*
  carbon emissions vs. GDP, *105*, 106
  and degrowth ideology, 106
  energy sources, 46, *47*
  impact of Messmer Plan, *164*
  oil crisis of 1970s, 163n
  and promotion of nuclear energy, 204
  and public perception of nuclear power, *182*, 189
  shift to nuclear power, 46–47
  and small modular reactor development, 197
  and spent nuclear fuel storage, 88
Friends of the Earth, 109, 151, 172–73, 181
Frisch, Otto, 4–5
fuel rods, *32*, *36*, 66n, *91*
Fukushima (Daichi)
  and antinuclear sentiment, 160–61
  cause of accident, 62n
  impact on Japan's energy policy, 183
  long-term effects, 74
  nuclear disaster described, 66–68
  and public perception of nuclear power, 142
  and relative dangers of energy sources, 63
  and small modular reactor development, 195
  and tritium production, 83–84

Fukushima (Daini), 62n
fusion, 30, *30*, 30n, 134, 215–16

Gabon, 53–54, 91
Gale, Robert, 73–74
gas peaker plants, 153n
gas stoves, 101–2
Gates, Bill, 196
GE Hitachi, 196
gender gap in support for nuclear, 185–88, *187*
General Mining Act, 75
geological repositories, 91. *See also* spent nuclear fuel
Georgia (state), 190–91
geothermal power
  and capacity factor of energy sources, *40*
  as clean energy source, 115–16
  and definition of renewable energy, 110
  and electricity production/consumption, 26, *26*
  and global energy sources, *116*
Germany
  and antinuclear sentiment, 68, 158–63, *162*, 170
  carbon emission levels, *159*
  and early fission research, 8–9, 15, 199
  Energiewende program, 158–59, 158n
  and energy diet concept, 120
  energy sources, 161, *162*
  and origins of antinuclear movement, 154
  radon gas spas, 71
germ theory, 13
Ghana, 190
global energy diet, *124*
global warming, 31, 52–53, 152, 200, 207, 216
Good Energy Collective, 185
Google, *192*, 192–93, 196
grassroots campaigns, xxi, 152
The Greatest Show on Earth (Dawkins), xvii
Greece, 73, 114
greenhouse gases
  advantages of nuclear power, 44, 47, 54, 58
  and author's pronuclear efforts, xxi
  and biomass fuels, 213
  and carbon footprint, 214
  and climate change, 25
  and closure of nuclear facilities, 149
  defined, 216
  and definition of renewable energy, 110, 217
  and energy diet concept, 122
  and fossil fuels, 215
  and home heating sources, 148
  and net zero goals, 41n
  and Ontario's energy sources, 45
  See also carbon emissions
Green Party (Germany), 160
Greenpeace, 145–46, 154, 174, 206
Green Planet Energy, 154
Grimes, 179

habitat loss, 43
Hahn, Otto, 3–5, 8, 30, 199
Hawking, Stephen, xv
HBO, 72–73
heat
  advantages of nuclear power, 44–45
  and causes of Chernobyl disaster, 65
  and creation of fossil fuels, 215
  and definition of energy, 215
  and desalination, 57
  and energy diet concept, 122
  and energy infrastructure in Brazil, xvii
  and fission process, 30, *35*, 215
  and geothermal, 115–16, 115n
  and global energy sources, 27
  and greenhouse gases, 25, 216
  historical background of power sources, 97
  industrial uses of high heat, 44, 163, 191n, 196
  and life cycle impact of energy sources, 55
  and natural nuclear reactors, 53
  and nuclear accidents, 147
  and origin of fossil fuels, 24
  and power generation process, 31, 35–37, *37*
  and promotion of nuclear energy, 206
  and radioisotope thermoelectric generators, 177n
  and solar power, 112
  sources of home heating, 148
  See also global warming

high-temperature gas-cooled reactors, 191, 191n
Hinkley Point C nuclear power plant, 154
Hiroshima bombing, 9, 200
Hitler, Adolf, 9
Hoff, Heather, 174–75, 179
Hollande, François, 165
Hollywood, xix, 85, 157
Homo sapiens, 21
horizontal drilling, 195
hot springs, 115
Human Development Index, 99, *99*
hydrogen
    and fusion, 215–16
    industrial production of, 44, 163, 191n
    and "proWindgas," 154
    and tritium production, 81–83, 81n, 218
hydropower
    Banqiao Dam disaster, 61, 63, 67–68
    and capacity factor of energy sources, *40*
    as clean energy source, 114–15
    deaths associated with power sources, *64*
    and electricity production/consumption, 26, *26*
    and energy diet concept, 117–19, 122, 128–29, 130
    and France's energy sources, *47*
    and global energy sources, *23*, *27*, *116*, *124*
    and Iceland's energy sources, *126*
    and Japan's energy sources, *184*
    and life cycle impact of energy sources, *56*
    and Ontario's energy sources, 45, *46*, *128*
    and relative dangers of energy sources, 63
    renewable energy described, 110–11
    and resource requirements of energy sources, *43*
    and U.S.'s energy sources, 38, *39*, *125*

Iceland, *126*
"the ick," 218
Idaho National Laboratory, 196
India, *182*
Indian Point, 149

Indigenous communities, 76, 77, 142
Industrial Revolution, 7, 22–24
inequality, 11
Inflation Reduction Act, 189
in-situ recovery, 79
Intergovernmental Panel on Climate Change, 52, 185
International Atomic Energy Agency, 83, 138–39
International Energy Agency, 163, 185, 193
International Monetary Fund, 51
International Renewable Energy Agency, 93
iron production, 24
Irwin, William, 150
ISODOPE, xx–xxi, 60, 171, 207, 209
isotopes
    and atomic structure, 34
    described, 81n
    detection of Chernobyl disaster, 65
    radioisotope thermoelectric generators, 177, 177n
    and research reactors, 32n
    tritium, 81–84, *84*, 218
Italy, *182*

Jackson, Ronny, 102
Japan
    atomic bomb attacks on, 216
    causes of Fukushima disaster, 66
    energy sources, *184*
    maintenance of damaged reactors, 83
    new nuclear facilities, 168
    nuclear attacks on, 10
    and public perception of nuclear power, *182*
    radiation exposure from Fukushima, 62–63
    restart of nuclear facilities, 183
jobs
    and author's pronuclear efforts, 60
    economic impact of nuclear power, 148–49
    and potential of nuclear power, 14
    and social benefits of transition to nuclear, 50–51, 58
joules, 69n
just transition, 50–51

Kairos Power, 192
Kaku, Michio, 177
Kean, Sam, 54
Kennedy, John F., 142–43
King's College, 150
Korean War, 10

labor-saving benefits of electricity, 97–102, *98*
Last Energy, 196
leaded gasoline, 144
Levi Strauss company, 55
life cycle emissions, 55–56, *56*
light water reactors, 32
LILCO, 152, *152*
linear no-threshold model, 71
lithium, 77–78, 120
lithium mining, 78
Los Angeles, California, 207
Lovering, Jessica, 185–87
Lovins, Amory, 108–10, 111

Macron, Emmanuel, 165
mutual assured destruction, 9–10
magnets, 6, 33, 114, 114n
Mangyshlak Nuclear Power Plant, 57
Manhattan Project, 9, 133, 137, 216
March for Environmental Hope, 174
Marshall Islands, 142
Marx, Karl, 96
Massachusetts, 149
Massachusetts Institute of Technology, 43, 57, 176, 179, 211
McKinsey & Company, 104
mechanical power, 22–23, 22n, 218. *See also* hydropower; turbines; wind power
media image of nuclear energy, xix–xx. *See also* public perceptions of nuclear power
Meitner, Lise, 4–5, 8–9
memes, xiii, 90, *90*, 96, 102–3
mercury, 24, 31, 88–89, 194
Merkel, Angela, 160
Messmer Plan, 163–64, *164*
Meta, 196
methane
and capacity factor of energy sources, *40*
and carbon footprint, 214
and deaths associated with power sources, *64*
and definition of fossil fuels, 215
and electricity production/consumption, *26*
and France's energy sources, *47*
and global energy sources, *23, 27, 124*
and greenhouse gases, 216
health effects of emissions, 24–25
and history of energy sources, 23–27, 28
and Japan's energy sources, *184*
and life cycle impact of energy sources, *56*
and mined resources, *79*
and Ontario's energy sources, *46, 128*
and origins of antinuclear movement, 152, 153–54
and potential of nuclear power, 7, 11
and resource requirements of energy sources, *43, 80*
and sources of methane emissions, *49*
and Switzerland's energy sources, *127*
and U.S.'s energy sources, *125*
MGX, 189
microplastics, 89
microreactors, 194–97, 216
Microsoft, 191, 193, 196–97
Middle East, 109
military's association with nuclear, 10–11, 75–76, 137–38
millisievert measure, 69–70, 69n, 216
mining, 74–77, 77–81, 95
molten salt reactors, xviii, 177, 194, 194n. *See also* thorium reactors
Morgan Stanley, 189
Mothers for Nuclear, 174, 175
Munroe, Randall, 86
Musicians United for Safe Energy (MUSE), 148
Musk, Elon, 103

Nagasaki bombing, 9, 200
Nash, Graham, 147
National Council on Radiation Protection and Measurements, 71
natural gas, *46*, 101–2, 153n
natural reactor. *See* Oklo region (Gabon)

Index

Natural Resources Defense Council, 146, 151, 173–74
Navajo Nation, 76, 80
Nazis, 8–9, 15
Nelson, Mark, *151*, 175
Netherlands, 113
net zero, 41, 41n, 183–85
neutrons
   and atomic structure, 33–34
   defined, 216
   and fission process, 4n, 35, *35*, 214, 215
   and history of fission research, 3–4
   and spent nuclear fuel, 86n
   and structure of atoms, 213
   and tritium production, 81–82, 218
Newsom, Gavin, 175–76, 179
New York (state), 149, *149*
The New York Times, 74, 147
Nipomo Dunes, 172
nongovernmental organizations, 183, 217
Non-Nuclear Futures (Lovins), 109
nonprofits, pronuclear, 206
No Nukes concert, 1, 148, 173
Norway, *159*
NPT, 138. *See also* Treaty on the Non-Proliferation of Nuclear Weapons
nuclear electricity (term), xviii
nuclear power
   and capacity factor of energy sources, *40*
   contrasted with coal, *31*
   deaths associated with power sources, *64*
   and electricity production/consumption, *26*
   fission process described, 29–30
   and France's energy sources, *47*
   and Germany's energy sources, *162*
   and global energy sources, *23*, *27*, *116*, *124*
   and Japan's energy sources, *184*
   and life cycle impact of energy sources, *56*
   lifespan of nuclear plants, 48
   and Ontario's energy sources, *46*, *128*
   reliability of, 41
   and resource requirements of energy sources, 41–44, *43*, *44*, 56, *79*, *80*
   and sources of methane emissions, *49*
   spent nuclear fuel, 59, 84–93, *85*, *87*, *88*, *91*, 150, 203, 217
   and Switzerland's energy sources, *127*
   and U.S.'s energy sources, *39*, *125*
Nuclear Regulatory Commission, 180, 196
nuclear semiotics, 90
nuclear weapons, 9–10, 129, 137–38, 141–43, 145–46, 145n, 156, 200

Oil Heat Institute, 152
oil power and resources
   and California politics, 173
   and carbon footprint concept, 101, 214
   deaths associated with power sources, *64*
   and electricity production/consumption, *26*
   and energy density of power sources, *42*
   and energy diet concept, 118, 122
   and France's energy sources, 46, *47*, 163–64
   and global energy sources, *23*, *27*, *124*
   and historical background of nuclear, 134
   and history of energy sources, 23–27, 28
   and Japan's energy sources, *184*
   and mined resources, 78
   oil crisis of 1970s, 104, 109, 163, 163n, 167
   and Ontario's energy sources, *46*
   and origins of antinuclear movement, 109, 151–52
   and potential of nuclear power, 7, 11, 199
   and resource requirements of energy sources, 41–42, *79*
   and small modular reactor development, 194–95
   and sources of methane emissions, *49*
   and Switzerland's energy sources, *127*
   and U.S.'s energy sources, *125*
Oklo (company), 196
Oklo region (Gabon), 53–54, 91
Onkalo deep geological repository, 91
Ontario, 45–46, *46*, *128*
OpenAI, 189
Oppenheimer, J. Robert, 9
Oregon State University, 84
organic-cooled reactors, 194, 194n
Ortiz-Wines, Paris, *151*, 175, 178

Our World in Data, *23, 27, 39, 42, 47, 64, 116, 123, 124, 125, 126, 127, 159, 164*
oxygen, 54n

Pacific Gas & Electric (PG&E), 172–73, 180
Palisades Fire, 207
Palo Verde, 57
Panama Canal, 194
Paris Agreement, 52–53
Partial Test Ban Treaty, 143, 145
particulate matter, 47, 148
Pennsylvania, 149
Persians, 113
Peru, 77
pesticides, 138n, 143–44
phossy jaw, 136
pitchblende, 75
plastics, 89
plutonium, 4n, 138n
Poland, 51, *159*, 189
Politico, 160
pollution, 11. *See also* carbon emissions
Popular Science, 137
population growth, 173
Porco, Carolyn, xviii, 177–79
Porter, David, 192
positive void coefficient, 65
potassium, 69n, 115
poverty, 11, 98, 107
pragmatic optimism, 13
#PrayForAmazonas, xiii
pressurized water reactor, *36*. *See also* reactors
pronuclear activism, 171–97, 203–7
propaganda, 101
protons, 33–34, 213, 217, 218
public perceptions of nuclear power
  American sentiment over time, *181*
  and author's pronuclear efforts, xix, 59–60
  countries compared, *182*
  and energy diet concept, 129
  gender gap in, 185–88, *187*
  and misconceptions of environmental movement, 200–201
  and proximity to nuclear facilities, 181–82
  and secrecy around nuclear research, 138–41
  See also antinuclear activism; pronuclear activism

racism, 187–88
Radiant Energy Group, 175
radiation and radioactivity
  and antinuclear sentiment, 160
  associated with coal power, 72, 72n
  and Chernobyl nuclear accident, 61–65
  and containment structures, 36, *36*, 36n, 214
  and cooling systems for nuclear plants, 57
  defined, 217
  and gender gap in support for nuclear, 188
  and geothermal power, 115
  and historical background of nuclear, 133–36
  and history of uranium mining, 75
  millisievert measure, 69–70, 69n, 216
  and nuclear meltdowns, 66n
  and nuclear power generation, 37
  and nuclear test bans, 145n
  and origins of antinuclear movement, 146–47
  physics of, 33–34
  and public perception of nuclear power, 59, 142–43
  radioisotope thermoelectric generators, 177, 177n
  and radon exposure, 77
  and relative dangers of energy sources, 68–74
  relative radiation levels, *70*
  and secrecy of nuclear research, 138
  and small modular reactor development, 194
  and spent nuclear fuel storage, 59, 84–93, *85*, *87*, *88*, *91*, 150, 203
  and tritium, 81–84, *84*, 218
  See also isotopes
radioisotopes. *See* isotopes; radiation and radioactivity
radium, 133–36
Radium Ore Revigator, 134–35, *135*

Index

radon gas, 69, 71, 76
Raitt, Bonnie, 147
Ramsar, Iran, 71
Ravikant, Naval, 13
reactors
   Aurora reactor design, 196
   core of, *32*, 35–37, *36*, *37*, *38*, 57, 81, 83, 86, 194n
   light water reactors, 32
   molten salt, xviii, 177, 194, 194n
   in nature, 53–54
   organic-cooled reactors, 194, 194n
   pressurized water reactors, *36*
   research reactors, 32n
   small modular reactors, 191–92, 193–97, 216
   thorium, xviii, 34, 115, 177, 194n
   regulation of energy sources, 67, 166–69, 180, 196
reliability of nuclear energy, 41
renewable energy
   and antinuclear sentiment, 163
   background of concept, 108–12
   and biomass power, 213
   and capacity factor of energy sources, 40
   defined, 217
   dirty vs. clean energies, *110*
   and energy diet concept, 120, 122, 130
   and France's energy sources, 165
   and Germany's energy sources, *162*
   and global energy sources, *23*, *27*, *116*
   and Iceland's energy sources, *126*
   and Japan's energy sources, *184*
   and mined resources, 77
   and origins of antinuclear movement, 153–54, 157
   and promotion of nuclear energy, 205
   and shifting sentiment on nuclear, 173
   and supplemental gas peaker plants, 153n
   and Switzerland's energy sources, *127*
   and waste levels of power sources, 93
reprocessing of nuclear waste, 92
repurposing fossil fuel plants, 49–50
retirement of nuclear plants. *See* decommissioning
rewilding, 66
Ring of Fire, 116

Rio Grande do Sul, Brazil, xiv
The Rise of Nuclear Fear (Weart), 133–34
Ritchie, Hannah, 78
rockets, nuclear, 7–8
Roosevelt, Franklin D., 9
Rosalía, xx
Russia, 9, 154, 158n, 161, 168, *182*, 196

sail power, 113
salaries, 51
San Luis Obispo, California, 172–73, 178, *180*, 181. *See also* Diablo Canyon
Sarkozy, Nicolas, 160–61
Save Clean Energy (SCE), 178–79
Save Diablo Canyon, 178, *180*
Schreurs, Miranda, 160
Schröder, Gerhard, 161
Science, 108
Science History Institute, 54
secrecy around nuclear research, 137–38, 141
Shellenberger, Michael, 173–74
Shippingport Atomic Power Station, 141
Sierra Club, 109, 152, 172, 174
Silent Spring (Carson), 143
silicon, 113n
Silicon Valley, 191
The Simpsons (television series), 37, 150–51, 157, 165
The Simpsons and Philosophy (Irwin), 150
Siri, William, 172
small modular reactors, 191–92, 193–97, 216
smog, 6
social media, xiii, xv, 60, 155, 204–5
SoftBank, 189
solar power
   and antinuclear sentiment, 161–62, 169
   and capacity factor of energy sources, 40, 40–41, 213
   as clean energy source, 112–13
   deaths associated with power sources, *64*
   declining cost of, 190–91
   and definition of renewable energy, 110–11
   and electricity production/consumption, 26, *26*

solar power (cont.)
   and energy diet concept, 117, 118–21, 128–29, 130
   and France's energy sources, *47*, 165
   and gender gap in support for nuclear, 186
   and Germany's *Energiewende* program, 158, 158n
   and global energy sources, *23*, *27*, *116*, *124*
   how solar panels work, 113n
   and life cycle impact of energy sources, 56
   and mined resources, 78
   and Ontario's energy sources, *46*, *128*
   and origins of antinuclear movement, 152, 154
   panels, 113n
   and potential of nuclear power, 7
   and promotion of nuclear energy, 205
   and relative dangers of energy sources, 63
   and reliability of clean power sources, 115
   and resource requirements of energy sources, 42–44, 42n, *43*, *44*, *80*, 179, 202
   and shifting sentiment on nuclear, 175
   and small modular reactor development, 195
   and social benefits of transition to nuclear, 50–51
   as source of most energy, 20–21
   and sources of methane emissions, *49*
   and supplemental gas peaker plants, 153n
   and Switzerland's energy sources, *127*
   and U.S.'s energy sources, *39*, *125*
   waste from, 93, *94*
   and waste levels of power sources, 92–93
Sørensen, Bent, 108
South Korea, 168, *182*, 183, 197
Soviet Union, 64–65, 142–43. *See also* Chernobyl nuclear accident
space requirements of energy sources, 41–44
Spain, 182
spent nuclear fuel, 59, 84–93, *85*, *87*, *88*, *91*, 150, 203, 217

Springsteen, Bruce, 148
Stand Up for Nuclear, 175
Stanford University, 43, 57, 120, 176, 179
Stargate Project, 189
steam
   geothermal power, 115n
   historical background of power sources, 97
   and industrial heat, 196
   and Industrial Revolution, 23
   and nuclear accidents, 147
   and process of power generation, 31, 35, *37*, 39, 58
   and turbines, 31, *37*, *38*, 115n, 196, 218
St. Lucie Unit 2, 168–69
Strassman, Fritz, 3, 5, 8, 30, 199
strong nuclear force, 33
submarines, 41, 138, 194
Susquehanna nuclear plant, 191
Sweden
   carbon emission of European countries, *159*
   carbon emissions vs. GDP, *105*, 106
   and Chernobyl disaster, 65
   and degrowth ideology, 106
   and history of fission research, 4
   reliance on nuclear, 121–22
   and shifting sentiment on nuclear, 183
   and World War II, 8
Switzerland, *127*
Szilard, Leo, 9

Taylor, James, 148
Technical University of Munich, 160
TED, xx
The Telegraph, 189
TEPCO, 83
terawatt-hour (TWh), 63, 63n, 218
Texas, 101–2, 120, 196
thorium reactors, xviii, 34, 115, 177, 194n
Three Mile Island
   and antinuclear sentiment, 146–48, 159–60, 167, 169
   impact on Swedish nuclear, 121
   and relative radiation levels, 70
   and tech industry's demand for power, 191, 191n
thyroid cancer, 74

time-saving benefits of energy, 22
transmission lines, 50
transuranic elements, 4n
Treaty on the Non-Proliferation of Nuclear Weapons (NPT), 138
tritium, 81–84, *84*, 218
Trump, Donald J., 189
tsunamis, 66–67. See also Fukushima (Daichi)
turbines
   capacity factor of energy sources, 40
   defined, 218
   and energy diet concept, 117, 121, 128, 130
   and France's energy sources, 165
   and geothermal power, 115n
   and hydropower, 114–15
   material resources associated with, 111
   and potential of nuclear power, 7
   and resource requirements of energy sources, 44, 202–3
   and steam power, 31, *37*, *38*, 115n, 196
   and waste levels of power sources, 93
   and wind power, 41, 114n
Twitter, xiii
Typhoon Nina, 61

Ukraine, 122, 161, 183
UN Framework Convention on Climate Change, 52n
UN General Assembly, 138
United Arab Emirates, 168, *182*, 183
United Kingdom, 143, *182*
United Nations' Conference of the Parties, 52, 52n, 183
United Nations Economic Commission for Europe, 54
United Nations Scientific Committee on the Effects of Atomic Radiation, 62
United States
   and antinuclear sentiment, 165–69
   and capacity factor of energy sources, *40*
   carbon emissions vs. GDP, *105*, 106
   energy sources, *38*, 38–39, *39*, *125*
   and nuclear weapons, 141–43
   oil crisis of 1970s, 163n
   and public perception of nuclear power, *182*
   and small modular reactors, 193–97

University of California, Los Angeles, 73
University of Chicago, 5–6
University of Oxford, 78
uranium
   and atomic structure, 34
   and definition of renewable energy, 110
   and fission process, 4n, 35, 216
   fuel pellets, *32*, 33, *42*, 86
   and history of fission research, 3–4
   mining, 74–77, 77–81, 95
   and natural nuclear reactors, 53–54
   number of protons in, 217
   and process of nuclear power generation, 58
   and reliability of clean power sources, 115
   and resource requirements of energy sources, 42
   and small modular reactor development, 194
   and spent nuclear fuel, 86
Urban, Tim, 20
U.S. Army, 194
U.S. Department of Energy, 41, 41n, 49, 51, 189
U.S. Navy, 194
utopianism, 14, 134–35

variability of energy sources, 120. See also capacity factor
Vietnam War, 10
Vogtle Electric Generating Plant, 167–68, 190–91, 195
volunteerism, 206–7
voting, xxi, 155, 175, 180–81, 205

wages, 51
Washington Examiner, 186
waste. See spent nuclear fuel
watch factories, 136
water resources and quality
   and American energy consumption, 102–3
   and author's environmental awakening, xv
   and Chernobyl nuclear accident, 61–63
   and definition of renewable energy, 217
   and degrowth ideology, 100

water resources (*cont.*)
  desalination, 44
  and energy diet concept, 117, 119, 129
  and function of turbines, 218
  geothermal power, 115n
  historical background of power sources, 97
  history of water power, 114
  and hydropower electricity production, 115
  and life cycle impact of energy sources, 55–57
  and mined resources, 76, 79
  and natural nuclear reactors, 53–54
  and nuclear power generation, 31–33, 35–37, 58
  and nuclear test bans, 145n
  and origins of antinuclear movement, 144–45, 157
  radium spas, 134–35
  and small modular reactor development, 194
  and spent nuclear fuel storage, 86, 92
  and steam turbines, 37, 38
  and tritium production, 81–84, 84
  water quality, 6, 81–84, 84
  wind-powered pumps, 113
  See also hydropower; light water reactors; steam
waterwheels, 22, 28, 114. *See also* hydropower
watts, 63n, 218
Weart, Spencer R., 133–34
weather patterns, xxi, 25, 39, 115, 200. *See also* climate change; global warming
Westinghouse, 196
What If? (Munroe), 86
white phosphorus, 136
wildfires, 207
wind power
  and capacity factor of energy sources, 39–41, 40
  as clean energy source, 113–14
  deaths associated with power sources, 64

declining cost of, 190–91
and definition of renewable energy, 110–11, 217
and electricity production/consumption, 26, 26
and energy diet concept, 118–21, 128–29, 130
and France's energy sources, 47, 165
and gender gap in support for nuclear, 186
and Germany's *Energiewende* program, 158, 158n
and global energy sources, 23, 27, 116, 124
and life cycle impact of energy sources, 56
marginal cost declines, 195
and mined resources, 78
and Ontario's energy sources, 46, 128
and origins of antinuclear movement, 154
and potential of nuclear power, 7
and promotion of nuclear energy, 205
and resource requirements of energy sources, 43, 44, 80, 202–3
and social benefits of transition to nuclear, 50–51
and sources of methane emissions, 49
and Switzerland's energy sources, 127
and U.S.'s energy sources, 39, 125
vs. hydropower, 114–15
and waste levels of power sources, 93
wind turbines described, 114n
World Health Organization (WHO), 62, 82–83
World Nuclear Association, 184, 188–89
World War II, 8–9, 75–76, 133, 138
Wyhl, Germany, 159
Wyoming, 51

X-energy, 191, 196
XKCD comic, 86
X-rays, 69, 70, 71

YouTube, 97

Zaitz, Kristin, 174–75, 179